# ENDOR~~SEMENTS~~

In being a "tuning fork, connecting humans and animals", author Anne-Frans Van Vliet elucidates energy healing modalities via clear case examples, and the transformative power of her work. In so doing, she brings awareness and hope to our relationship with animals, and its greater impact.

Claudia Six, PhD, USA
Clinical Sexologist & Relationship Coach
Author of Erotic Integrity: How to be True to Yourself Sexually

Anne-Frans Van Vliet's, **Animal Energy Therapy Project** is an essential contribution to the growing scientific field of Trans Species Communication and Healing. This book gives practical methods of both HTA (healing touch for animals) and scientific evidence for the sentience of a spectrum of mammalian lifeforms on the planet. I am also inspired where *Animal Energy Therapy Project* sits within the larger context and urgent imperative of Homo sapiens: to recognize the sacredness and intelligence, and deep interconnectivity of all life forms on the planet before it's too late. Whether you are an animal lover who wants to give expanded care to other species or someone who is looking to keep pace with an expanding and essential world view that will help us truly shape the social policies and spiritual consciousness of the 21st century, read this book ASAP.

Sarah Drew, USA
Author of Gaia Codex

Reading this book will both open your heart and activate your mind. It is as inspiring and heartfelt as it is informative and enlightening. Anne-Frans's passion and love for the animal kingdom is clear and undoubtedly assists her in effectively healing and transforming the lives of the animals she treats. By sharing her process, she reveals the power of combining research, practiced and proven energy healing techniques, ancient wisdom, modern medicine and Divine Love. Not only is this a recipe for healing our animals, but a profound tool for healing humanity

Beki Crowell, USA
Soul Artist, Vibrational Healer
Author of Bare Beauty: My Journey of Awakening

The world needs this book! We now have a valid research on the importance of animal energy therapy practices with various animal species impacting their health and well-being. Anne-Frans uses her purpose on this earth and impacts our collective divine responsibility to take care of all off God's creatures. And this begins at home.

Michelle Sevigny, Canada
Founder of Dogsafe Canine First Aid

Through her memoir, I can feel Anne-Frans's fascination for life. It is amazing how Anne-Frans discovered during her different trips, her own peak of sensitivity through the energy of the elephant, the power of connecting with such a wonderful animal and to discover the possibility of sensory communication and healing with one another. This book is all about emotions, spirituality, senses, perception and full of love and joy.

Darwin Angulo, Mexico
Dog Coach

# ANIMAL ENERGY THERAPY PROJECT

One Woman's Journey
Healing Animals
through Energy Work

Anne-Frans Van Vliet

**Animal Energy Therapy Project**

For information, email annefrans@annefrans.com
www.annefrans.com

Published by:

ISBN-13: 978-0-578-56368-8

# DEDICATION

In dedication to my mother and father, who were strong Dutch role models. Their teamwork throughout their long-lasting marriage propelled us to seize life and make it meaningful and exceptional. My father's brilliant career as a foreign news correspondent and author guided me to find my voice. My mother and grandfather planted the roots for my work with animals, promoting healing, wellness, and longevity.

# CONTENTS

# ACKNOWLEDGMENTS

To my incredible husband, Tom Van Dyck, whose visionary approach to environmental and human rights encouraged me to follow my passion. To my warm and supportive parents, my sister, Eleanne, and brother, Nick, who have my back no matter where my adventures take me. To my inspirational mother-in-law, who wrote two books after the age of 75. Through the strong encouragement from Alicia Dunams, I decided that the voice of animals needed to be shared. Thank you for your belief in my capabilities and continuous support from your incredible team of editors and designers throughout this journey. I want to send my love and appreciation to my close friends and the people who helped me manifest the vision and supported me throughout the process of birthing this book: Beki Crowell, Cynthia Simon, Lauren Levine, Robin Shapiro, Claudia Six, Carol Komitor, Steven A. Nielson, Bobbi Cohen, my teachers and mentors, and my friends and support team in India. Finally, I want to thank all the animals that have given me unconditional love and happiness. Thank you.

# FOREWORD

Animal Energy Therapy Project provides simplicity and understanding of how our animals work through the eyes and presence of Anne-Frans Van Vliet. This information travels beyond the individual animal, expanding to cultural differences, species and the planetary reactions the animals provide. Her humanitarian presence exceeds most caring humans and her voice speaks through her heart. She defines the science behind her work, the nuances the animals show and the love she holds for all species.

The information provided in this book will help all of us to better understand the inner workings of our animals, how they affect our lives, how we affect theirs and how without animals, humans will not survive.

The stories offered here are heart-felt and encouraging as they demonstrate how people can connect with the animals beyond their daily care. These stories will help us to better understand and bond with our own animals.

I have witnessed firsthand, Anne-Frans' work and how her passion drives all that she does. This passion is expressed within the pages of this book demonstrating the collection of knowledge Anne-Frans holds and her life-long passion for animals. Her education, bio-energetic understanding, one-on-one sessions and her research allows us a peek into the window of who she is and the difference that can be made when desire is present.

An easy read, Animal Energy Therapy Project will be an asset to all who love animals and their people.

—Carol Komitor, Healing Touch for Animals®, Founder

# INTRODUCTION

*The eternal being, as it lives in us, also lives
in every animal.*

~ Arthur Schopenhauer

A s human beings, we can build meaningful lives when we connect with one another and with other species in authentic ways. It is in our best interests, as well as the best interests of the world around us, that we keep our planet green and in optimal working order, and we strive to be conscious stewards in protecting every living being from harm, extinction, pain, and suffering. Each of us has the possibility and the opportunity to discover what we can do individually and collectively to contribute, give back, and take a stand as activists, role models, and leaders.

In this book, you will encounter stories of several beautiful animal characters I had the pleasure to work with in Jaipur, India. Through my experiences, I will present the powerful effects of integrating animal energy therapy techniques in the daily lives of animals and the incredibly transformational impact the animals had on me. I encourage everyone, through these inspiring stories about healing, deepened understanding, and connectedness with other species, to recognize that each of us has the capability to create a fulfilling and purposeful existence when we connect with our animal kingdom and nature. These connections enrich our lives and raise our vibration to higher frequencies and instinctual alignment with the Divine and our Universe.

I have been an animal lover, advocate, and guardian for as long as I can remember. I cannot recall a time when I wasn't concerned about the health, well-being, and longevity of the animal species around the globe. When growing up in the Netherlands, we heard about animals in remote places and their struggles for survival. We lived around farm animals and dogs and cats, but unless I pursued a career as a veterinarian, spending quality or quantifiable time with animals wouldn't be feasible. As a result, any time spent with animals would be nothing more than my hobby.

In college, I started as a pre-med major, but I soon realized that I was intrigued by psychology and learning about the cognitive brain and behavior patterns. It wasn't until I had finished my Master's Degree in Public Relations and Business Communications that it dawned on me that I needed to follow my heart and pursue a career where I could actually work with animals on a full-time basis, not just on my days off. By that time, I had become more in tune with who I was, what made me the happiest, and what career path I truly desired. Every step I had taken was pulling me in another direction, toward my purpose and goals in this lifetime.

However, in order to pursue my passion and prepare myself to work with animals in the way I wanted, I knew I needed more education in the animal behavioral sciences and embarked on a new adventure. Years later, having worked daily with animals for more than 15 years, I can feel that the developments that transpired have played a pivotal role in what my deeper purpose is shaping up to be. My purpose is to be a tuning fork, connecting humans and animals to each other and creating a bridge of understanding between all species in order for us to coexist in harmony and take better care of one another and the planet.

In this book, I share some of the experiences that shaped me and touched my soul. I have had a front row seat witnessing the profound impact these experiences have had on the beautiful animals I treated.

## Current Historical Developments

Over the past 30 years, there has been a rise and incredible interest by animal behaviorists, ecologists, conservation biologists, evolutionary scientists, theorists, physicists, neurobiologists, and more in studying and documenting their research findings about the consciousness and emotional sensitivity of animal species and their expansive brain capabilities. Animal experts agree that most species have a conscious brain. "Cognition relates to the kind of information an organism gathers and how it processes and applies this information," writes Frans de Waal in *Are We Smart Enough to Know How Smart Animals Are* (2016, 69). Our current research confirms that almost all animals experience emotions and feelings, just like humans.

Yes, animals are quite similar to humans in many ways. Humans and all other species are affected by the environment they live in and environmental stressors that surround them. Studies continue to prove that external stimuli are the leading cause for disease and chronic illnesses, such as cancer and heart disease, as well as the leading cause for behavioral and psychological abnormalities, such as anxieties and depression.

In order to improve the health and wellbeing of all species, in my opinion, we need to optimize, harmonize, and balance the energy system of all beings. This is the desired route for everyone living on this planet, which naturally suggests that we should utilize all the tools and techniques at our disposal, which have been crafted, refined, and practiced throughout centuries by healers, doctors, and masters all over the world.

## Today

Veterinary medicine is the most accepted animal healing practice available. However, energy therapy techniques have become more readily accepted and practiced since we have highly-skilled animal healing professionals around the world. It is now possible to become educated in a variety of specific modalities of animal energy therapy practices. In doing so, we can guide not only our human selves to be in optimal balance, and can also assist our animal counterparts in achieving the same balance.

Through listening, respecting, and simply being present with our animals, we are able to develop a deeper connection, understanding, and more fulfilling relationship with them. With higher degrees of understanding and knowledge, we will act with more vigor as stewards and call others forth to be advocates for all species and our environment.

As we evolve together and co-exist on this planet, it is important to preserve the health and longevity of all species for reproductive purposes, especially those that have existed on this planet longer than homo sapiens and are at risk of extinction or subject to animal cruelty. Many national and international NGOs (Non-Government Organizations) are demanding that governments worldwide ban animal cruelty and provide more care and protection for all animals. Conscious governmental ruling, strict policies, and a continuous supply of financial aid are imperative for the survival of the many creatures we treasure, aim to preserve, and want future generations to witness.

Due to increases in the human population, we need to continue efforts that promote the well-being and longevity of wildlife, domesticated wildlife, and animals that live on farms, swim in oceans and bodies of water, fly in the sky, and reside in our homes. More sanctuaries need to be built to house both wild and domesticated populations. "By steadfast voicing our concerns, upholding our beliefs, and taking action, we can bring about change. Once a sufficient number of people take action, we can make changes that benefit all animals and the Earth (Goodall and Bekoff 2002, 120)."

## *My Goal*

I wrote this book to shine a bright light on and bring awareness to our God-given duty. In our own ways, we have every opportunity to be brave and bold enough to stand outside of our human comfort zones and create positive and profound impacts to improve the planet and all its living forms. For example, I decided to branch out from my work with dogs, cats, horses, and animals on farms and traveled to India to work with elephants. I dove in head first to experience life with a species that I had not worked with yet. I was ready to listen, learn from, and create a deeper understanding and relationship with the domesticated Asian elephant and become a part of their immediate ecosystem. I longed to be close, where I could observe them in their daily life and communicate with them in their environment. I wanted to feel their emotions and observe their conscious minds, experience their magnificence, and perceive the world through their eyes. I had a craving to feel and experience what it was like to be in their clan, and I deeply wanted to contribute and give back. Toward that end, I knew I would be able to make a positive contribution to their lives by giving them Healing Touch for Animals® energy therapy.

It is my goal to raise global awareness about the benefits of consistently integrating animal energy therapy practices with species all over the planet. The purpose of my work, while listening to the animal, is to integrate modalities of healing and provide each of them with the most optimal and holistic approach of health and healing possible, behaviorally, spiritually, psychologically, and physically.

## Chapter 1

# ANIMAL ENERGY MEDICINE & HEALING TOUCH FOR ANIMALS® (HTA)

*The greatness of a nation and its moral progress can be judged
by the way its animals are treated.*

~ Mahatma Gandhi

## History of Energy Medicine

Some of the first indications of healing therapies and medicine date back 60,000 years ago, when homo sapiens used plant medicine for ingestion, to place on wounds, and to stimulate the olfactory system by inhalation. Research shows that we have existed on this planet for 300,000 years. Chinese healers crafted Traditional Chinese Medicine 2,000-plus years ago when they identified 12 major meridians, pathways of energy in the body that are charged by energy vortexes called dantian centers. These meridians link all limbs and organs together in an intricate physical network. The Chinese medical perspective relies upon the interrelation between heaven and earth, our animal and human physical and emotional body with nature and the environment. This theology is articulated through the dynamic relations with yin and yang, the Five Elements theory and qi.

When this energy field is strong and vibrant, the entire organism maintains health and balance. When it is weakened and disturbed by external stimuli, the body becomes increasingly subject to weakness, sickness, and sometimes death. It is our natural intelligence that possesses a magnetic and genetic pull, directing living organisms toward health for reproduction, passing on genes, raising families, and survival. Any deviation from this innate attraction to grounded wellness creates ill health.

Over centuries, healers using energy medicine discovered that they were able to positively influence the body's health by working with its energy field. As we advance, we elevate consciousness, and we increase frequency and vibration levels in all living beings.

In other parts of the world, Ayurvedic and Tibetan Medicine has been a fundamental system of healing. These systems of medicine recognize the energy system comprising of Chakra centers, energy wheels, or vortexes as main engines of energy connecting to Meridian pathways.

For over 2,000 years, certain cultures have relied on energy medicine for the livelihood of their livestock and farm animals. The Asian population has used Traditional Chinese Medicine, and the Indian population has focused on Ayurvedic and Tibetan Medicine for centuries. The integration of various styles of healing was readily accepted and available, since Veterinarian medicine did not exist and was uncommon. Western human medicine was recognized in the 5th century B.C. in Greece, and it has evolved over the centuries. Veterinarian medicine was not invented until 1761.

Today, due to globalization and the vast integration of peoples and cultures, our innate ways and practices have blended together. Overall, the human population is more aware and open to believing in and accepting alternative healing methods. We live collectively on this planet, and animal illness, disease, and behavioral challenges are on the rise. It doesn't only affect them; it affects all of us and the planet. With continued animal activism efforts, the desire for optimal health and longevity for all species continues to push us forward to grow, change, and evolve.

## *My Overall Experience*

In my case studies and in the healing that I experienced with my work in India, I noticed a common thread. Even though each animal had a different persona, a different way of being, and had experienced different frictions and blessings in their lives, integrating animal energy therapy and holistic approaches that showed love and care proved to positively impact improved health and the lessening of behavioral issues.

## *Healing Touch for Animals®*

Carol Komitor, the founder of Healing Touch for Animals® (HTA), was involved with Healing Touch Program™ and studying and working in the human energy side of medicine. Ever since she was a child, she has had a strong connection with animals and the ability to see energy fields. She believed that animals would benefit from energy therapies that were created especially for them. In an interview with Carol, she explained that due to their heightened sensitivity and "pixelated" energy field, she designed techniques that promoted healing and calibration of the energy system that are animal-specific. She mentions that "on an instinctual level, they know when we mean them no harm; they are driven by unconditional love, they have no agenda and are able to receive the energy work without putting up boundaries. They bring us in, they let us help them, and they are able to adjust energetic changes in their own health and well-being, whether it is on an emotional level or on a physical or even instinctual level. (*Interview May 2019*)."

## The Body's Energy System

In order to fully grasp energy therapy and its benefits, it is important to understand the body's energy system. All living beings are made up of energy, which means that when there is life, there is an energy system. The energy system of living beings is made up of seven main Chakra centers. These Chakra centers and their locations directly correlate with functioning organs and glands. Meridian pathways transport energy to all systems within the body, such as the circulatory, lymphatic, muscular, and nervous systems. The energy field that surrounds us outside our body is called the Aura. It is important to recognize that environmental stressors and stimuli are processed by our internal energy field, and these vibrations pass through the energy field as they travel through our body. This energy connects our physical being to every living being and to the Universe, Divine, and Mother Earth.

Carol states in her interview, "Our energy field collects the data of our life experiences and stores it. This field can hold energetic baggage from past trauma, disease, and/or illness. Clearing this congestion assists the body in the healing process. If the energy field is not clear, the energy system cannot balance, and health is compromised. The energy field changes according to what is going on around the animal or person, if there is health, joy, and happiness, the energy field will expand. If there is illness, injury, or trauma, the energy field tends to constrict, creating a lack of movement and energy flow."

In comparison to the animal energy field, the human energy field is quite different. The human energy field consists of four distinct layers that extend about four to six feet past our physical bodies. These four layers are: 1) the physical body, 2) the emotional body, 3) the mental body, and 4) the spiritual body connecting to God and Universe. Attaining and maintaining a consistent balance between these layers is vital for health and well-being.

Through evolution, human beings have a lesser instinctual capability, and we use our more developed logical brain to make decisions. This means that we need to be even more aware of all the intricacies that can affect us and use both our intellectual and instinctual abilities for optimal well-being.

The energy system of the animal consists of one large instinctual and pixelated energy field. In comparison with humans, animals use the edge of their energy field to read their environment and make instinctual decisions based on the information felt from the outside inward. Their instinct allows them to determine what is naturally good or harmful for them. The energetic field of the animal extends out at least ten times the size of their physical body, which is vastly larger than our human energy field. The animal instinctual field is pixelated, not layered, and will not collect energetic debris or toxins between the layers, as it does with humans. The animal is able to clean and release energetic blockages from traumas, injuries, and illnesses with greater ease due to their pixelated and fully integrated energy field. This is another reason why animals are so receptive to energy medicine and why it is so beneficial for them.

Through HTA techniques, "actual molecules of energy are stimulated, flow through the body and increase the life force of that body. The pixels in the pixelated energy field of animals are able to line up and the body becomes supported, with a better sense of balance and creating a sense of stability and strength within their own energy fields so that they can release debris and toxins and move forward with their present life. They are able to 'right' themselves and bring that energy field into a consistent flow, and supporting optimal health (*Interview May 2019*)."

HTA is a holistic approach that uses energy medicine and intention to influence and facilitate the health and well-being of all animals. The techniques used to restore harmony and balance the animal's energy system, while providing physical, emotional, mental, behavioral, and instinctual stability. When the energy system is stabilized, a natural physiological relaxation response sets in and supports the immune system, which, in turn, promotes well-being by encouraging the healing process as the body recalibrates and returns to homeostasis. Homeostasis is the natural preferred state of being, and it is the most optimal and desired state for the body to function at its highest level.

TREATMENT SESSION: In each session, exact techniques of optimal benefit to the animal are incorporated. As practitioners, we execute approximately three to six techniques in a one-hour-plus session. The techniques used are based on our intake assessment findings and initial observations.

OBSERVATION & ASSESSMENT: We observe the animal and the caregiver informs us of any symptoms that are of concern and out of alignment, such as behavioral abnormalities, illnesses, injuries, and past traumas. We use HTA with healthy animals to maintain the functioning of their systems at optimal levels, as well as with animals that have illnesses, injuries, or behavioral inconsistencies.

The assessment is a way for the practitioner to measure the integrity of the energetic body, which is made up of Chakra centers that are in constant motion. From these motion centers, energy travels throughout the body to all organs and tissues via the Meridians. If an energy center is blocked or compromised, the energy moving forth is not as vibrant and, therefore, provides less than efficient "juice" to the whole body. This can cause physical and physiological breakdowns in the body.

Using various tools, we are able to assess whether the energy centers are flowing openly or not. The vibration from the Chakra energy centers will transmit to these tools. Using them, we assess the Chakra centers and the Hara. Energy travels everywhere in different wavelengths and vibrations, and even though it cannot be seen by the naked eye, it can be felt and assessed for vitality. Rest assured, this is a reliable science, not guesswork.

The variation of HTA techniques used can be categorized into certain "sense" categories. We use our hands for touch therapy. We place our hands directly or indirectly on various places of the body and hold that position for continuous and direct energy transfer. We also move our hands to allow the energy to flow over and into the whole body. Due to energy transfer, we can perform these techniques from a distance, as well. In some of the healing techniques, the practitioner stands farther from the animal, opening up to the animal's larger energetic field and encompassing the animal's broad energy field.

We use tuning forks for sound therapy. We place them directly on certain parts of the body, either on energetically congested areas or on a specific Chakra center, stimulating vibrational movement. In this therapy, the animal listens to the sound waves that we strike in deliberate patterns and sequential orders.

Tuning fork sound therapy was developed in 1998 in the school of "Inner Sound." This is an original system of sound therapy and music for human beings. Carol effectively adapted this methodology for use with animals. Compared to humans, animals have greater sound sensitivity, which is why sound therapy is a component of the different variations of energy therapies that exist for animals. Sounds can be very effective in helping the animal ground and be present, allowing for a deeper healing experience.

In HTA, we use an OM tuning fork set. When these forks are struck in a certain sequence, the vibrational tones combine to create the universal sound of OM, which is called "a pure 5th interval." The OM tuner correlates to the earth's natural frequency and is recognized as complete "homeostasis" within the body. These tuners create movement within the energetic body to break up and clear congestion and help the body return to a balanced state or equilibrium, where it is able to function optimally. The 5th interval is thought to be the most important interval when used in sound healing. The grounding effect empowers the person or animal to open their heart center, allowing the body to relax and, in turn, recalibrate itself.

We also use essential oils for Limbic and aromatherapy. We use high-quality essential oils by Young Living™ to enhance the effectiveness of the techniques used. Essential oils (EO's) can be applied topically on the body, or they can be administered via inhalation and/or ingestion. Regardless how they are administered, EO's work directly with the Limbic system of the brain. This part of the brain houses the amygdala, as well as the hypothalamus, and is located in the lower part of the brain, which is often referred to as the "reptile brain." All animals and humans have this part of the brain, which is connected to the processes that control physiological functions.

EO's provide the energetic vibration and natural properties derived from pants, trees, and flowers to support the body's energetic system. Each plant holds molecular properties, known as constituents, which have vibrations specific only to that plant. By stimulating the Limbic system, the body sends signals to the brain, which triggers the body physiologically to respond by tapping into the constituents of the plant and naturally blending together in union.

## *Indicators of Relaxation in an Animal*

Physical signs of relaxation are indicators that the body is responding to the energy work that is taking place or that the HTA session is concluded. The body will exhibit universal signs of relaxation, such as lowering of the head; softening and closing of the eyes and eyelids and/or the lips and mouth; gurgling of the stomach; breathing or sighing deeply; settling down energetically; the release of intestinal gasses, urine, and excrement; stretching different parts of the body; licking and chewing, etc.

When the body is in a relaxed state, it has a chance to recalibrate, reset, and "right" itself. Studies have shown that relaxation 1) improves oxygen flow to organs, 2) strengthens muscles and bones, 3) decreases the risk of diseases, treats and prevents various diseases, 4) heals injuries and illness faster, 5) reduces muscle injury, arthritis, and inflammation, 6) re-energizes the brain and balances hormonal levels, increasing serotonin and the release of endorphins, 7) helps the animal to understand appropriate behavior, 8) strengthens the animal-human bond, 9) reduces stress and anxiety, 10) supports animals through physical and emotional trauma, abuse, and grief, 11) builds a solid foundation for cancer patients, 12) develops confidence for training and competition, and 13) energetically supports the animal through the end-of-life transition.

HTA is being practiced in all parts of the world and with all species. Its effectiveness continues to be proven, and increasingly more people are learning and becoming a part of the HTA community. Carol Komitor reminds us that, as energy practitioners, we need to "remain grounded and connected to our spiritual selves in order to be able to facilitate someone else. Our self-healing is imperative if we are going to do this work because we can only heal subliminal, unless we have healed ourselves (*Interview May 2019*)."

## Chapter 2

# ELEPHANTS & INDIA

*If they live, they love. If they love, they care.*
*If they care, they feel.*

~ Anthony Douglas Williams

## *The Elephant and Her Evolution*

Elephants have evolved and existed on this planet for over 55 billion years. In contrast, homo sapiens have only existed on this planet for 300,000 years. That is only one-half of one percent of the existence of the majestic elephant. Elephants have endured many changes that have taken place during their lifetime and their overall existence, including constantly being threatened by the increase of the human population, the human-animal conflict, and domination of and capitalization by humans.

Even though only a few elephant species remain today, they have survived. At one time, elephants lived all over the world, even in Rome and the Mediterranean. Over the past 55 billion years, about 300 elephant species have roamed the earth; however, now only 3 species (two African and one Asian) remain. Not only have their species and numbers decreased, but their habitat has, as well. India is home to about 50 to 60 percent of the remaining elephant population, and about 20 percent are domesticated; the others remain wild.

Elephants have one of the largest brains of all species. Their cerebral cortex is larger than ours. Just like the human brain, most of an elephant's brain development takes place after birth, states Carl Safina in his book, *Beyond Words, What Animals Think and Feel* (2015, 43). "Elephants are the largest land animals, and sperm whales the largest toothed mammals in our animal kingdom. Of all animals, these two have the largest brain size, which are also, in both species, large relative to their body size. Brain size gives a rough measure of mental flexibility, compared with intelligence, and large mammalian brains are associated with social complexity. Sperm whales and elephants have both evolved into a lifestyle that resonates with ours, living long lives, producing few offspring, and holding high investments with each offspring," as stated by Katy Payne in her book, *Silent Thunder: In the Presence of Elephants* (1998, 118).

Elephants have the capability of expressing the primary emotions that Charles Darwin's research proved and which he publicized in the mid-19th century, such as fear, anger, happiness, and sadness. These primary emotions are easy to understand since they are triggered first in response to an environmental situation and are usually instinctual based. Elephants, however, are also able to express secondary emotions. These secondary emotions precede the primary reaction and are more complex. Some examples of secondary emotions are jealousy, joy, and shame. In *The Emotional Lives of Animals,* Marc Bekoff remarks that elephants are said to be "the poster species for animal emotions, since they display so many emotions so deeply (2007, 41)."

Elephants' memories are superior to humans since their survival has depended upon remembering copious amounts of information. Evolution has shaped these elephants over thousands of years to remain instinctually grounded and extremely capable of natural survival. The elephant's memory enhances its ability to identify potential threats and to use their past experiences as a survival mechanism. Due to the increase of the human population, human-animal conflict, and the need for dominance and control, wild animals are at extreme risk of extinction.

Elephants' memories are not only expansive, but the elephant's brain exhibits behaviors that were once thought to be in the sole domain of humans. For example, elephants show signs of post-traumatic stress disorder (PTSD) when they have experienced trauma. Research has shown that elephants can actually have nightmares. A report from an elephant orphanage in Kenya mentions that African baby elephants wake up screaming in the middle of the night, remembering that they witnessed their elephant family members being brutally assassinated and tusks being cut out of their bodies by poachers (Bekoff 2007, 45). Cindy Moss speculates that elephants do have a concept of death in *When Elephants Weep* (Masson and McCarthy 1995, 96). Animals, just like humans, will show their emotions by using their body language and actions. "All we have to do is look, listen and smell. Their faces, their eyes, and the way in which they carry themselves can be used to make strong inferences about what they are feeling. Changes in muscle tone, posture, gait, facial expression, eye size and gaze, vocalizations, and odors (pheromones), singly and together, all indicate emotional responses to certain situations" states Marc Bekoff in *The Emotional Lives of Animals* (2007, 45).

Elephants and other species will also produce tears when experiencing a wide range of emotions, from grief and pain all the way to joy and pleasure. When I was working with the elephants in Jaipur, I, indeed, also experienced and witnessed their joy, energy

release, and relaxation in the form of tears and water gland stimu-
lation in relation to the energy work they were receiving.

"To understand an elephant, one must be 'anthropomorphic,'
because elephants are emotionally identical to ourselves. They are
sensitive, have emotions, and show feelings. They grieve and mourn
the loss of a loved one just as deeply as do we, and their capacity
for love is humbling," stated Daphne Sheldrick as documented by
Carl Safina in *Beyond Words: What Animals Think and Feel* (2015, 73).

Elephants are able to communicate to other elephants up to
13 km away without us hearing any part of their communication.
"Elephants' low-frequency rumbles create waves not only through
the air but also across the ground. Elephants can hear and feel these
rumbles inaudible to humans over distances of several miles. Their
great sensitivity to low frequencies derives through ear structure,
bone conduction, and special nerve endings that make their toes,
feet, and trunk tip extremely sensitive to vibration, meaning that
part of elephant vocal communication is sent through the ground
and received through their feet. Millions of species communicate
using scents, gestures, postures, hormones and pheromones, touch,
glances, and sounds (Safina 2015, 78-79)."

Elephants live within very strong social networks with other
elephants. More often than males, females seek out and select one
another and connect as a strong female friendship group. Each of
these groups has one leader, who is called the matriarch. Usually,
the matriarch is the oldest and wisest female of the group. On the
other hand, the male elephant is quite independent and will travel
back and forth from clan to clan, both female and male, sometimes residing with only a small group of males as they mature
and learn from one another.

Elephants show an incredible variety of love emotions, not
just love on an instinctual level but love on an intellectual level,
as well. "Love is a feeling. It motivates behaviors such as feeding

and protection … The capacity for love evolved because emotional bonding and parental care increase reproduction (Safina 2015, 54)."

It has never been completely understood if animals, just like humans, actually know who they are in their physical body. Human beings have the capability to distinguish one from the other, recognizing each physical being as someone other than themselves. When we look in a mirror, we recognize that image portrayed as being a reflection of ourselves. Research of Asian elephants has found that some of them will recognize their own image in a mirror (de Waal 2016,18). This has been a huge discovery as this, indeed, connects more closely the elephant species to that of the human species.

In conclusion, elephants are incredibly social in nature, have a high degree of intellect, memory, a sense of self, and the ability to learn and live side by side with humans. "They are caring beings and show a high degree of loyalty, bonding, affiliation, and cooperation (Safina 2015, 97)."

Champa 1 and Shavitry

## *Elephants in Jaipur, India & My Research Process*

Lord Ganesh is of one of the most famous Indian Deities still worshipped today. Lord Ganesh has a human boy body with an elephant head. He is considered a god of wisdom and intelligence. While the elephant is part of Lord Ganesh's body, elephants have also been a part of Indian culture and history for centuries. The bond between humans and elephants in India through historical revolution has always existed in a palpable way.

In Jaipur, the capital of Rajasthan, about 120 female elephants remain. Jaipur continues to be a well visited tourist destination and elephant attraction due to elephants being a status symbol that is utilized in many Hindu and political ceremonies and festivals dating back to the 1600s. Amber Fort, a monument and former palace during those early years, is now a UNESCO (United Nations Educational, Scientific, and Cultural Organization) world heritage site, where elephants remain to be a popular staple. In 2018, the local Rajasthan government finished building the Elephant Village, an elephant sanctuary for domesticated elephants, housing working and retired elephants, their Mahouts (caregivers) and their families.

I was able to work with a population of female elephants through a nonprofit organization called Volunteer with India (VWI). This organization allows tourists access to about 10 of the female elephants residing in Jaipur. The purpose of this organization is to provide volunteer opportunities in several of their programs. The programs consist of working with street children at an orphanage, teaching English, teaching young women sewing, the henna trade, and feeding, cleaning, and caring for elephants.

With this domesticated elephant population, there is a special connection between elephant and human. The elephants have come to rely on humans for food and a high degree of care 24 hours a day. Over the years, many animal rights activists have forced state and national governments to put in place stricter rules and measures in relation to the care of these domesticated elephants. At VWI, those rules are upheld and regulations are followed.

I had the opportunity to observe and live side by side with the elephants and the Mahouts for several months and take note of the close bond and strong relations between the two.

## *My Research Process*

After my successful first visit in India, where I volunteered with VWI in the elephant program, I decided to return in October 2019 and conduct my explorative research study doing HTA energy therapy work, and I filmed and documented my findings. I visited and worked in the same environment that the elephants reside in, observing the elephants in their conditioned setting.

I worked with a population of eight female elephants and was able to conduct at least six full treatment sessions with each one during my stay. All eight elephants had different personalities and characteristics. Some exhibited behavioral nuances, and some of those nuances were more dramatic than others. Their nuances varied, as did the animals I worked with—I also had the pleasure of working with horses, goats, and dogs.

For each energy session, I intentionally only used a few previously selected energy therapy techniques because I wanted to take note of the value and benefits these specific techniques had with the elephants and be consistent with my protocols. The techniques I selected used touch, sound, and smell. I varied the order in which I used the selected techniques, based on my daily assessment findings with each elephant at the beginning of each session. The goal was to attain the most beneficial structure of techniques during each session.

I was interested in observing which technique provided optimal results specific to each animal by producing more noticeable immediate relaxation results, as well as continued long-term effects and changes in behavioral patterns. Each treatment session with each animal was filmed and recorded on paper, documenting the animals' immediate responses during the session and the changes that took place over the course of the month.

The EO's I selected to use were Sacred Frankincense, Frankincense, Palo Santo, Copaiba, Bergamot, Myrrh, and Ylang Ylang. These are the reasons I chose these specific oils.

*Sacred Frankincense* – This oil comes from a steam distillation process of the Boswellia sacra frankincense tree. This oil is regarded as the most highly-prized form of frankincense and is connected to a bible oil. This oil triggers a higher spiritual connection, in addition to providing many health benefits for conditions such as asthma, depression, allergies, colds, diarrhea, meningitis, respiratory problems, and ulcers. It is immune strengthening, amongst many more benefits. This oil can be inhaled, ingested, and absorbed through the skin.

*Frankincense* – This oil has been used since ancient times for sacred and medical purposes. The oil is sourced from the resin of the Boswellia cartei or sacra tree. This oil has a sweet, woody, earthy and uplifting scent and a broad range of uses. It is used for grounding and spiritual connectedness, creating a safe and comforting environment, reducing heart rate, stress levels, and blood pressure, and increasing immune function. It can also benefit coughs, colds, indigestion, ulcers, depression, arthritis, infection, inflammation, laryngitis, multiple sclerosis, and Parkinson's disease, and has many more benefits. It can be inhaled, ingested, and absorbed through the skin.

**Palo Santo "Holy Wood"** – This oil is steam-distilled from wood chips of dead heartwood found in Ecuador. This clean and inspiring fragrance has anti-infection, anti-tumor, and immune strengthening properties. It is also used for muscle pain, acute rheumatism, nervousness, and spiritual awareness. For centuries, this oil has been used by indigenous peoples of the Amazon in rituals and ceremonies for purification and cleansing of toxic energy and misfortune. This oil can be inhaled and applied topically on the skin.

**Copaiba** – This oil is obtained from a steam distillation from Oleoresin, a flowering plant from Brazil and Ecuador. It contains high levels of beta-caryophyllene and has a sweet aroma. It creates a calming atmosphere. It has strong antiviral and antibacterial qualities and stimulates the circulatory and pulmonary systems. It is great for healing wounds and cuts, internal infections, anxiety, muscle aches and pain, respiratory problems, urinary tract problems, stomach ulcers, tuberculosis, and tumors, plus much more. It can be inhaled, absorbed through the skin, and ingested.

**Bergamot** – This oil is extracted or pressed from the rind or peel of the Rutaceae citrus fruit. With effervescent citrus tart aroma, it also has a sweet and relaxing quality. It is used as an antibacterial ointment for strep and staph infections, is anti-inflammatory, antiparasitic, aids digestion, and is uplifting. This oil can be inhaled, ingested, and applied topically.

*Myrrh* – This oil has been used since ancient times and is popular in the Middle East and Mediterranean. It is also a bible oil. This oil comes from the steam distillation of the red-brown resin of trees in the gens Commiphora. It has been used for medical and sacred purposes. It has a bitter taste, yet an earthy and sweet aroma, and luxurious properties that are used in skin care that are calming and uplifting and bring a deeper sense of spirituality to daily life. It is believed to ease coughs and colds, soothe digestive discomfort, anti-tumor, astringent, and boost immunity. The compounds enhance health, including terpenoids, a class of chemicals with antioxidants and anti-inflammatory effects, alleviate pain, and promote healing. This oil can be inhaled, absorbed through the skin, and ingested.

*Ylang Ylang* – This oil has a sweet romantic aroma and is heart related. It is commonly used in aromatherapy. It is sourced from the flowers of Cananga odorata, a plant native to the Philippines, Madagascar, and Indonesia. One of the main components of this oil is linalool, a compound found to possess stress-reducing properties and increase a feeling of calmness. Other problems this oil is used for are anxiety, colds, sinus infection, headaches, fever, insomnia, muscle tension, pain, infection, intestinal problems, skin problems, inflammation, mood, and depression. This oil can be inhaled, ingested, and absorbed through the skin.

## Chapter 3

# RASHMA ~ Female horse

*Not all of us can do great things. But we can do small things with great love.*

~ Mother Teresa

### Day One

My first journey to India was in the fall of 2017. I was working toward fulfilling the requirements to earn my certification in HTA, and the certification process required me to work with a variety of different species in order to broaden my experiences. I decided that I wanted to work with domesticated wildlife. Travel has played a large part in my life. I had been craving a trip to India and decided this would be the perfect spot to do the work. It was a fantastic opportunity for me to travel and work with the animals. I love to connect with people and animals, experience different ways of life and cultures, and become one with another part of the world, as I listen, observe, and learn.

In India, I not only worked with elephants, but also with horses, goats, dogs, and humans. One of the horses I worked with was Rashma, a 25-year-old, small, white female horse. Rashma's guardian for the previous three months had been Atchou (or "animal mother"). He also took care of Bhitily, another female horse, and is the Mahout of several elephants at the elephant barn. Rashma's job was to be a part of the Hindu Indian wedding ceremony, where it is tradition for the groom to travel by horse.

On the very first day of my arrival at the elephant barn, where a lot of other species were housed, Atchou wanted me to meet his horses and take a closer look at Rashma, in particular, and see if I could help her. In his broken English, he told me, "she … problem." Not knowing what I was in for, I followed Atchou through a hole in the wall of the elephant barn to a dilapidated, dusty, hot yard that had a tiny tin roof in one corner, providing the residing animals with a little sun protection. Here, I met Rashma and her sister, Bhitily, who were surrounded by goats.

As I walked over to Rashma, I could see she was extremely thin. What shocked me most was that part of her back and hind quarters on the left and right side of her spine was covered with blistering wounds. I was stunned and taken aback to see a horse in such bad shape. It looked like she had been badly attacked and mutilated by another animal. The skin surrounding the open wounds was blistered, red, infected, and inflamed. Near the infected area, her white hair had fallen out; only fresh flesh was showing, and flies were feasting upon her open wounds, spreading their bacteria into her system. Her physical body looked frail and emaciated. I could see her ribcage and hip bones protruding, dangerously pushing her thin skin tight. She was skittish and hyper aware of her surroundings, including me. Right away, I could tell she was ungrounded, and her spirit felt weak and broken. This girl was in deep trouble. I feared she might not survive her massive injuries and the obvious infection that had overtaken her body.

I learned from Atchou that Rhasma had, indeed, been brutally attacked by a male horse a few days before, and he did not have enough money for a veterinarian to clean and treat Rashma. Atchou asked me if I could help her heal and get better. Without any hesitation, I said yes.

When I gently approached Rashma, she was overly alert and watchful of my every move with eyes that were huge and open wide. I slowly walked toward her and stood next to her, speaking in a very calm and even toned voice. My light touch made her shiver. Looking at the grave situation in front of me, it made sense that she also had deep internal bruising and was in tremendous pain. I knew I had to work quickly, for she was obviously seriously ill. My mind was racing as I tried to think of the best strategy for creating rapid healing. Observing the environment and the sweltering sun that was blazing down on Rashma, I first decided to move her to the shaded part of the dirt yard. I wanted her body focused on healing and conserving energy, not fighting the summer heat and blistering sunshine. When I touched her body, I could feel that some parts were transmitting higher degrees of heat than others; these hotter areas were also clammy to the touch and smelled rancid and infected.

## My Approach

When working with an animal, I approach each situation by looking at a cohesive whole, including the environmental factors surrounding the animal. I observe an animal in a holistic way so I can effectively integrate the most logical tools and skill set that I have to aid in healing. For Rashma, I decided to integrate Western medicine first aid care, add nutritional foods and supplements to enhance her overall strength, and conduct regular HTA sessions. As a canine first aid and CPR instructor, I have been able to tap into my skill set with many other animals in need and travelling with some basic first aid supplies has always come in handy.

I started by cleaning her wounds with filtered bottled water, which I followed with saline solution for extra disinfecting in and around the wounds. After this cleaning, I dried the wounded areas. I did not have antibiotics with me, but I always carry Copaiba essential oil, a popular and acclaimed high-frequency essential oil with viral and bacterial infection fighting agents. Using little drops, I spread the oil in circles around the massive wounds. I wanted the oil to penetrate into the skin and push out any internal infection, kickstarting the healing process. I learned this technique from Larry Brilliant.

On the way to India, I had listened to an inspiring audio book called *Sometimes Brilliant* by Larry Brilliant, the former director of Google.org, who now serves as Chair of the Skoll Global Threats Fund and is the 2006 TED prize winner. Together with a team of doctors from the United Nations, he was instrumental in helping to eradicate smallpox in India in the 70s. Due to a massive outbreak of smallpox, a shortage of vaccines, and the inability to contain the spread of the disease, his team came up with an incredibly smart defensive strategy, using their limited vaccine supply wisely. In their impressive plan, they encapsulated the infected people by vaccinating the uninfected human beings around them. The idea behind this plan was to build a contained perimeter and systematically vaccinate and stop the spread of this deadly virus. This ingenious system proved to be highly successful, and they were able to contain the smallpox outbreak and eradicate the spread of it throughout the Indian continent.

I was very taken by Larry's stunning story and his insightful ways of looking at the tremendously huge problem that faced a massive population in India and needed swift immediate care.

From him, I learned that I could encapsulate Rashma's massive wounds with Copaiba EO from the outside and have her ingest it through her drinking water so she could fight the massive infections inside her body.

## Day Two

The following day, I checked in with Rashma and Atchou. With a big smile, Atchou mentioned that Rashma had actually been able to lie down and sleep on her side for a few hours the previous night. He told me that she had not been able to lie down to rest since the attack, most likely due to the pain and discomfort of the internal bruising and inflammation. I was thrilled to hear this great news and knew if she was able to sleep and rest her body on the ground, she would heal much more quickly.

When I approached Rashma on the second day, she was more open to receive me, and her overall demeanor was brighter and more grounded than it had been when I first met her. She was happily grazing on some hay that hung in a graze bag on a hook and looked up at me with a gentle gaze and friendly acknowledgement. Looking into her soft eyes, I could feel that she had a sense of belonging and had regained some strength. I did notice, looking at her wounds, that there was fresh blood and fresh scabs were developing. When asked, Atchou mentioned that during that morning, she had been rubbing herself against the sides of the barn and gate. I knew this was also a sign of great progress. She was relieving the intense burning and itching she was experiencing, and, in doing so, she had reopened the wounds. Oftentimes when the body receives any sort of medical treatment or energy work, it seems to take a step back before moving forward into the healing phase. With my basic first aid knowledge, I knew that the burning and itching sensations were caused by internal toxins. Once the toxins release through open skin, the itching stops, and the wounds are able to scab and heal. This also meant that Rashma had responded to the energy therapy I had conducted with her the previous day and was turning an initial healing corner.

Working with Rashma was only part of the puzzle. It was imperative for me to establish a trusting professional relationship with Atchou. In doing so, I would be able to convince him that Rashma's healing depended on creating the best environment for her in order to promote rest and healing. I wanted to ensure that Rashma would physically, spiritually, and emotionally rebound quickly from the brutal attack. I needed regular access to her, which made it important that my bond and connection with Atchou was positive and that he was happy with my work and understood the environment he needed to create for her. I educated Atchou on patience and the time it could take for a natural healing process to complete its cycle. I had to make sure he understood that he needed to give her time to heal and secure his word that he would not force her to work until she was fully healed. Because Atchou was very poor, and Rhasma was a working horse that brought him income, changing his view from looking at the short-term gains over the long-term gains was not easy. We had different viewpoints, and to make matters even more complicated, we did not speak the same verbal language. I kept my part of the agreement by working to heal Rashma, hoping that he would keep his end of the deal in return.

## Transformation During the First Week

I integrated HTA, increasing her nutrition, feeding her bananas and larger doses of food, and giving her first aid care. I monitored her closely on a daily basis, observing her overall healing and behavior patterns. I did notice rapid daily changes taking place in her physical, emotional, and behavioral health. Not only were her wounds healing, but she also became more receptive to my touch and had a more joyful demeanor.

I continued to treat the wounds topically with Copaiba and gave her Copaiba to ingest and inhale. By using these methods, the vibrational qualities and molecular makeup of Copaiba could travel through the Limbic system and have a positive impact on her psychological state, as well as kill off any infections residing in her physical internal body. In addition to Copaiba, I used touch techniques such as Ultrasound, Laser, Vibrational Grooming, and Magnetic Clearing to focus on clearing resin and toxins from her hind quarters in between her root, sacral, and solar plexus Chakra centers. The internal bruising and trauma needed to be flooded with fresh, consistent energetic vibrations, promoting a continuous rapid healing process.

Another effective technique I used frequently with her is called Bridging. This technique creates a bridge, joining the heart and throat Chakra in a solid joint pathway, allowing her to express the unconditional love she holds within herself throughout her whole being constantly and consistently. Sending herself energetic waves of her own Divine love stimulated her wellness.

Her overall responses to the sessions were very gratifying. When I started working with her, she was aware and vigilant, but she portrayed a stiffness in her body posture, as well as a lack of energy, a sense of defeat, and overall dullness. As the sessions progressed, she became more relaxed and started to accept and adhere to my light touch. I could feel both energetic and physical shifts taking place continuously. The energy I felt inside the Chakras had more eminent determination. The work readily triggered her relaxation responses, such as eyes softening, head lowering, lips quivering, drooling, stomach gurgling, releasing gas and urine, and having bowel movements.

After the first week, I could tell that Rashma was transitioning and becoming a completely different horse. Her mood and attitude were much lighter, yet stronger. She was more grounded and energetic than she had been in the prior days, and she was now eating heartily and looked up to greet me every time I came to see her. As our relationship grew, she started to show trust and responded to my touch by leaning into me. She had a much higher energy vibration, and her pain levels were diminishing. Flies were no longer attracted to her wounded body parts, since the infections had cleared. I felt that we had accomplished a lot during the first week, and she was healing accordingly.

## HTA Energy Work During Week Two

On Monday afternoon, I was back at the elephant barn to check in with Rashma and Atchou. Atchou mentioned that she had been doing fine and there was not much new to report. However, I noticed he seemed a little agitated, perhaps impatient, and he even said he wanted her to heal faster. He asked what I could do to speed up the process. I could feel his anxiety, yet I also knew that natural healing takes time, especially after such a brutal attack. Again, I worried that he didn't understand the seriousness or the scope of her injuries. I wondered if he was aware of the incredible magnitude this incident had on Rashma. She had undergone some tremendous injuries, and it would simply take time for her to heal properly and fully. My fear was that he didn't grasp that concept and would put Rashma back to work too soon, which could cause a setback or a relapse.

Rashma was residing in her usual spot in the hot dirt yard next to the elephant barn, and I walked over to examine her. Her wounds looked like they were healing nicely and the skin around the wounds was less red and inflamed, more pinkish in color and not as puffy as they were the week before. She showed me recognition and gentle affection as I touched her in greeting. I continued to examine her body by running my hands over her to feel for hot spots and sensitivities, while I smelled her to see if I could detect any infections. Overall, her physical body was healing extremely well, but I did notice that her energy was low. I decided to give her some more bananas and some fresh water and conduct an HTA session with her. I knew that Rashma still had a lot of healing to undergo, and I had my work cut out for me in my last week with her.

During the assessment of this next session, her root, sacral, solar plexus, and crown Chakras were all closed. Her Hara was open. These were very different findings from the previous week, and I became concerned. At the end of the first week, her Chakras were open and flowing with full vigor. How had she taken such a leap backward during the weekend? I could not help thinking that Atchou had worked her by having human beings ride on her back and that she had not received the rest she still desperately needed. Atchou came over to watch me work, and when I asked him, he admitted that he had allowed her to be ridden. I was disappointed. Obviously, I had not done a proper job of convincing him that he needed to grant her sufficient time to fully heal. If given that time, she would heal much faster, and they would be able to work together again. I needed to emphasize that in a way which he could understand that Rashma was not going to heal overnight. There was no magic pill to give her; he and I needed to be on the same team and let the natural healing process come full circle. I let him know that, even though she looked better on the outside, she was still not strong enough to withstand people riding on her back and could easily take a turn for the worse if she was not taken care of properly.

After our brief talk, I continued my work with Rashma. I used my trusted Copaiba EO as a topical application on her wounds and Chakra centers, and then had her inhale the scent and gave her some to ingest through her hay and drinking water. Then, I proceeded with a technique called Chi Balance, providing a high-voltage energy transfer for Chakra balancing and stability. I followed this with another hands-on technique called Ultrasound, which I had already consistently used on her hind quarters throughout each of the previous sessions. I then conducted the Vibrational Grooming and Magnetic Clearing techniques. These specific techniques were used to loosen and release any tightness, inflammation, clogs, and debris that continued to reside in her body. I concluded the session with Etheric Heartbeat. As practitioners, we use this sacred technique at the end of sessions to envelope the animal in her own Divine light, unconditional love, and universal Divine power to strengthen and heal her. Her Chakra and Hara channels were all open and flowing consistently again at the closing assessment of the session.

## Transformations During Week Two

Her ability to lay down, sleep, and rest again during the first and second weeks had definitely helped her in the overall healing process. In addition, the extra nutrients and food supply strengthened her physical body and further aided in her healing process.

I also witnessed noticeable improvements in the healing of her wounds, and both the internal and external infections were no longer present. Every energy session conducted with her over the two-week period had stimulated her body into relaxation mode, which allowed for the opportunity of recovery as her immune system strengthened. In addition, her Chakra centers opened up rapidly during the second week of work as her body softened, absorbing the energy transfer with ease and desire.

At the end of the second week, it was time for me to leave and travel to other parts of India. I felt extremely satisfied with all the work I had been able to do with not only Rashma, but the other animals at the elephant barn, as well. I hoped Atchou would take better care of his horses and Rashma would continue to heal and become healthy and strong again. I was planning on a longer stay the following year as I had fallen in love with Jaipur and the animals that had become my friends. With that in mind and hoping that Atchou had taken my words to heart and would continue to let her heal, I was looking forward to seeing Rashma again in a year.

## Back in California

A few months after my return to California, I contacted the Mahouts to inquire about the animals at the elephant barn and see how Rashma was doing. It was exhilarating to hear the great news that Rashma had been doing well, was physically stronger, seemed happier, and her wounds were fully healed. I was so happy to hear that her hair had been growing back and her coat had more luster again.

In February 2018, however, I received a call and heard the terrible news that Rashma had not survived the long winter and had passed. I was heartbroken, angry, and terribly disappointed with myself and also with Atchou and her other caregivers. I felt that we had failed in our responsibility as a team to protect this amazing animal and keep her from death's door. After many inquiries, to this day, I sadly still do not understand the full story of what exactly caused her death.

Rashma's death inspired me to become even more involved and advocate for all animals. The day I learned of Rashma's passing, I made a solemn vow to increase my animal energy healing practice with animals all over the globe, continue to learn from masters and teachers, and become more vocal in expressing my message, thereby helping more animals. Today, I vow to share the impact of my work with others, as we all are capable of helping where and when we can. We are competent stewards for other species, while remaining respectful to other peoples, cultures, and beliefs, and together, we will continue to carve a path toward more understanding and the possibility of seamlessly coexisting on our earth.

## Chapter 4

# CHAMPA 1~ FEMALE ELEPHANT

*What you do makes a difference, and you have to decide what
kind of difference you want to make.*

~ Jane Goodall

### India, September 2018

Before I embarked on my adventurous journeys to India, I had not yet worked with animals that had eating disorders until I met Champa 1, a 43-year-old female elephant. Champa is a very picky girl and had chosen her one caregiver, Jamad, with scrutiny. She is the matriarch of a family of eight female elephants that reside together in a rugged, pink-toned barn made of stone and plaster, that had an open roof in the middle. This low building was quite unassuming, and one would never suspect that it housed eight beautiful female elephants.

From the hotel, it took me about 30 minutes every day to get there in a loud tuk-tuk, otherwise known as an Indian taxi. A tuk-tuk is a small, partly enclosed motorized mode of transportation that is quite popular in India. It can nimbly zig zag and zip through the chaotic Indian traffic that swarms over thoroughfares, narrow alleyways, and side roads, all of which are riddled with potholes and comprise varying degrees of pavement and dirt surfaces.

This journey from the city of Jaipur to the elephant barn is exhilarating to me every time. All my senses not only absorb the colorful nature, the loudness, and over-populated scenes that are Jaipur's backbone, but also the hot sun and the spicy air penetrating into my lungs and soul. I don't even mind the smell of pollution, as this is a part of India today, and the India that I have fallen in love with. On the early morning trips, once we arrived at the elephant barn, the warm Indian summer sun would rise over the hills in the distance, outlining the Hindu temple with its bells ringing perfectly in unison, accompanied by a chorus of prayer and chanting. This was breathtaking, and it silenced my being for a few moments. I felt full of bliss and gratitude to have the chance to be in India and be a part of the rhythms of living life with the elephants.

I was able to visit the elephants every afternoon five days a week, and I was able to see them twice a day two days a week. In the early mornings, we were there to wake the elephants, clean their stalls, and bathe and hydrate them before they went off for their morning tourism outings at Amber Fort. Every afternoon, I had the opportunity to do my research work and HTA with them.

The elephants stood in a row side by side, leaving a large pathway through the middle of the structure. Champa 1's position was at the entry of the barn, and she heard me coming from miles away. She was always the first to greet me when I walked in, slinging my camera, tripod, and gear bag over my shoulders as my eyes adjusted from the bright sunlight. She was blind in her left eye, and her right eye was infected and had impaired vision. I was informed that she could become fully blind in the near future.

But that wasn't her only ailment; she also had an eating disorder. Most elephants eat about 150-200 kg of food per day, and Champa 1 consumed about 200-250 kilograms of food per day. When Champa 1 did not receive enough food, she got quite temperamental and made sure everyone knew she was hungry. When tourists were around her, she could be extra moody and uninterested in connecting with them. This state of unbalance not only created anxiety for Champa 1, but also for the Mahouts.

Elephants can be ever so gentle and incredibly sweet and sensitive, but just like any species, they can also feel annoyed and show anger. Due to their incredible size, this can be quite dangerous, as you can imagine. Not only could Champa 1 be temperamental with humans, but also with her fellow female elephant clan. Like any other elephant that could pose a danger, she surely needed to be treated with respect and understanding.

During my initial assessment and observations of her, more questions started to form. What could be the root cause of her eating disorder? Was this behavioral abnormality attributed to her environment and the life she led daily as a working elephant? I knew she played a major role and participated in Hindu ceremonies, festivals, and tourism. For about three hours in the morning, she carried tourists around on her back, and during festivals, which occur regularly in Jaipur, she is dressed up and painted with bright colored nontoxic body paint. Were any of these things environmental stressors that were causing this behavioral disjoint?

Furthermore, I had other factors to consider and went through a checklist of questions. What are her current surroundings like? Is she taking on a behavior pattern or anxiety that is not hers to take on, but is perhaps present in one of her human counterparts, such as her Mahout? Could it be that her Mahout shows his emotions and stressors, and she picks up on these emotional waves?

Maybe Champa 1's past was difficult in some way, subjecting her to various traumas when she was growing up, such as a shortage of food. Did she witness and/or experience scarcity and abuse?

We also needed to consider the medical perspective. Was this eating disorder physical or perhaps genetic? It could mean that her internal organs were not functioning properly and weren't sufficiently retaining and processing the nutrients she was ingesting. If she wasn't absorbing the nutrients and minerals she was ingesting, then her body was craving more food, and her neurological brain was telling her to eat more in order to be satisfied. Her eating behavior could be attributed to all of the potential causes and more.

This situation was not an acute or short-term illness. This had been a part of her life for many years and was chronic in nature. Her physical body, as well as her energy field, was being weakened by this illness, and it affected her both behaviorally and physically. As an Animal Energy Practitioner, I wanted her to be able to identify and clear the underlying cause of her eating disorder and work to "right," re-energize, stimulate, and re-calibrate her energy system.

## Insights from Others

Carl Safina states, "Humans and elephants have nearly identical nervous and hormonal systems, senses, milk for our babies; we both show fear and aggression appropriate to the moment (2015, 19)." He further suggests that their behavior shows their awareness of what is going on around them. Christof Koch writes, "Whatever consciousness is … dogs, birds, and legions of other species have it… They, too, experience life (Safina 2015, 23)." They experience life, and they feel and show emotion. They feel just like human beings do. They might even feel in more ways than we do, and those feelings may be indescribable in our language. It is important to take note of this simple fact as we consider the relationship we have with our animals and the importance of listening to the animal and taking in the full spectrum when we navigate the best ways to ensure optimal health, wellness, and longevity for each animal.

"The part of the brain called the limbic system, which is thought to mediate emotion, is one of the most phylogenetically ancient parts of the human brain, so much so that is sometimes called 'the reptile brain.' From a purely physical standpoint, it would be a biological miracle if humans were the only animals to feel. To understand animals, it is essential to understand what they feel. When an animal is hurt in a way that would hurt a person, it generally reacts much as a person would. It cries out, it gets away, then examines or favors the affected part, and withdraws and rests. Veterinarians do not doubt that wounded animals feel pain and use analgesics and anesthetics in their practice (Masson and McCarthy 1995, 16, 23, 29)." Young elephants have even been known to die due to a broken heart if they have been separated from their friends and family.

Champa 1

## *HTA Energy Therapy Sessions*

As a professional, I start each session by setting the energy space around the animal and myself, aligning it to my highest good intentions. My intent is to make sure I do not hold on to or have any attachments to the outcome or results of the work. The recipient of the work will utilize the energy in the manner their spirit and body see fit.

It is always extremely exciting and overwhelming to be in the presence of these elephants. Their incredible beauty and sense of knowing radiates and sings through my spine. Showing up vulnerable, humble, and authentic as a full being with highest integrity and connected to mother earth, the Divine, and the Universe is a must. As practitioner and recipient, sharing this work together is extremely sacred and sensitive and needs to be treated with utmost respect.

I was able to conduct six full sessions with Champa 1 during my month's stay. For all sessions, my assessments showed a compromise in the Hara system. This energetic line connects the physical being with the spiritual and the earth.

From the information gained at the start of the sessions, I designed the layout of techniques to use in each session. For every session, I changed the layout and varied the techniques if I felt it was needed. I used a few essential oils (Sacred Frankincense, Frankincense, Palo Santo, and Myrrh) consistently with her that, when tested energetically, provided the highest benefit.

After each session, the energy transmitted during the session continued to transform her body's energy system, and I could start to see the changes take place. I could even feel the shifts in energy within her energetic body during each session, and I could see the physical changes, such as relaxation indicators, take shape.

## *Observations*

During the days between the sessions, I carved out extra time on a daily basis to sit with her, observing her magnificent beauty and communicating and snuggling with her. She loved to snuggle up close and have one-on-one girl time. There were many moments when I would just sit in the brick opening leading from the elephant barn to the dirt patch where the horses and goats mainly reside and be present with Champa 1. She would come close to me, lower her head, reach her brow toward me, and gently touch my forehead, the lower part of her trunk resting on the ground in a half circle. We would pause there for some time with our eyes closed in a peaceful embrace. The emotions we shared during these magical moments were of pure bliss and quite indescribable. When she would pull away and open her eyes, she wouldn't break her gaze with me. I stroked her trunk with my hands and kissed her often, smelling the sweetness of grass coming from her breath and feeling the cool firm pink skin of her trunk against my lips. There is nothing more exhilarating to me than those moments. It was as if time stopped completely. What an extremely lucky girl I was to receive such tremendous love on a daily basis, such a gift from such a majestic lady.

The results that took place during the month with Champa 1 were instrumental in many ways. I believed that the beginning process of recalibrating her energy system would continue to strengthen her full energetic body over time. I learned from her as we bonded closely and came together as trusted friends. We cherished the impact and joy we both gained from understanding one another in our different species' ways. This loving energy work allowed us to communicate together on a deeply vibrational level.

Due to the chronic nature of Champa 1's situation, this type of energy work needed to be conducted on a regular weekly basis in order for her to "right" through the past traumas, learn to remain balanced in her daily environment, and navigate through her vision impairment and painful eye infection that were holding her in the space of disability. In order to provide continuous consistency of energy care, I taught the Mahouts some general techniques and left essential oils for them to use regularly with all the elephants in their care.

As research has shown, directed energy transfer can be as simple as spending quality time with an animal, stroking her, and just being present with her. You and I have the opportunity to partake in these grounding and soothing practices on a daily basis with the animals that surround us. Given the benefits for both parties, there should be nothing stopping us.

## Chapter 5

# MAMA & FOSTER BABY
# ~ FEMALE GOATS

*Like a shepherd, He will tend His flock. In His arms, He will gather the lambs and carry them in His bosom; He will gently lead the nursing ewes.*

~ Isaiah 40:11

### Individuality Between All Animals

I am always amazed by the close connection that I feel with diverse species all over the world, yet each individual animal has their own story. The eight elephants I worked with all have different personalities and character traits, yet they are of the same species, the Asian elephant. Due to these individual variations, no energy therapy session is ever the same. While the sessions and techniques are different, all animals are affected by the energy work; they show different relaxation responses, prefer certain techniques over others, and will process their own balance, centering, and healing at their own time.

In addition to the elephants living in the elephant barn, I was in daily contact with many other species. Horses, goats, pigeons, dogs, and cows reside and filter in and out constantly. When I was there, it was the end of September and also baby season. As a result, there was an adorable little baby goat walking in and around the elephant barn.

During the third week of my stay, one lactating mama goat suddenly passed away. She was bitten by a poisonous snake that was living in the weeds next to the elephant barn. I didn't know that I was working in such close proximity to a venomous snake and was upset that this had occurred. The beloved baby goat was now motherless and needed regular nursing. While there were other female goats around, none were lactating and would be able to raise her. For the time being, Atchou, the main Mahout of the farm animals, was going to need to bottle feed this little baby goat on a regular basis.

One early afternoon, as we were feeding the baby goat a bottle of milk together, he mentioned that he needed to buy a new lactating mama goat to care for the baby. I looked at him in astonishment and asked if that was possible. Can one just replace a mama goat that easily? How is this new mama goat going to accept her foster baby and nurse her? Can one just plug in a new lactating mama goat and the natural process of caregiving would then proceed smoothly? To me, a lot more than just a simple replacement would be needed, but I was interested in seeing how Atchou would handle this situation. It didn't take long before I would find out.

## Insights from Others

On an ongoing basis, I read a plethora of animal stories, articles, and books, and I watch animal films. I have an insatiable appetite to learn everything there is to know about various animal species. I know that animals in the same species need to feel love and affection for each other in order to care for and protect each other. "When we feel love for our own babies, it's instinctive, not intellectual. Situations produce hormones, and hormones produce feelings. It might be as automatic as the letdown of milk – but we feel it as love. Love is a feeling. It motivates behaviors such as feeding and protection ... The capacity for love evolved because emotional bonding and parental care increase reproduction... Emotions that motivate us to erase a distance, to protect, to care for things, to participate... Love isn't one thing, and human love isn't all identical in quality or intensity... The word that labels ours also labels theirs. Love, as they say, is a many splendored thing (Safina 2015), 54, 72)." When animals take care of one another and their young babies, there is a strong presentation of love, affection, and empathy. "Human empathy is a critically important capacity, one that holds entire societies together and connects us with those whom we love and care about. It is far more fundamental to survival, knowing what others know... We share this with all animals... It is essential for reproduction, since mammalian mothers need to be sensitive to the emotional states of their offspring, when they are cold, hungry, or in danger. Empathy is a biological imperative (de Waal 2016, 132)."

Mama & Foster Baby

## *Back at the Elephant Barn*

The following day when I arrived at the elephant barn, I heard some exciting news. A new lactating mama goat had been purchased, and Atchou took me to the dusty yard behind the elephant barn to meet her. I immediately noticed that she looked uneasy and disconnected. I sat for a while, observing her demeanor and behavior, and I saw that she was out of sorts. This new environment did not feel like home to her, and she was definitely not attracted to the orphaned baby goat. Surprisingly, the baby goat kept wanting to nurse from her, but the mama kept kicking her away and walking off in another direction. It dawned on me that the mama goat needed to receive some HTA energy work to ground her and connect her to her new surroundings and her new foster baby. In cases like these, there is really no other source of medicine or training to speed up the process of establishing an energetic bond between two strangers. Energy therapy is the only proven way.

This was a new situation and one I had never encountered before. I was thrilled and eager to conduct an energy session with the mama goat and witness the changes that would take place. What actually transpired was beyond anything I could have imagined. While I was working with the mama goat, the baby goat was so attracted to the exchange in energy between the stranger and myself that she wanted in on it, too. I was able to conduct a dual session with both of them simultaneously and bridge the two of them together. I wanted to initiate a natural instinctual bond between the mama and baby and add emotion to that instinctual bond. I was grounding the new mama goat into her new environment and her new role as a mother, facilitating the connection between the two.

In the beginning of the session, it was a little difficult for the mama to remain still. She was fidgety and constantly moved away from me and then moved back toward me. I had to be patient and, without any expectations, continue the energetic transfer and allow her time to come back to me, although I purposely stayed within close reach so I could continue the work by touching her physically. In addition, the little baby kept coming up to her, but she was not interested in her in the slightest. After the first 15 minutes or so, the mama goat started to settle into the energy session. She stopped moving back and forth and voluntarily came to stand next to me. Her breathing deepened dramatically, and she settled in even more. As she stood there, I witnessed her release the tension and anxiety that she had displayed, and she began to open up, relax, and receive this energy transfer.

Later that night, I watched the film footage of that energy session, and I noticed that the mama was showing a presence and an acceptance, and she was drawn to the energy work. For the first part of the session, my focus was on working solely with the mama goat. However, as the mama goat settled in, and with the baby goat staying next to us, I was able to include the baby goat, using touch to create an energetic connection between the two. With one hand placed on the baby and the other on the mama, I was the bridge that brought the two together. This special moment lasted for some time and was naturally broken by the baby jumping down from the doorsill where I was sitting. The moment was very special, and it was a turning point in their relationship. While I hadn't known what to expect, I had not anticipated or planned this scenario and was happy with the way the session had naturally taken shape.

## *Observations*

The following afternoon, I observed both of their behaviors, and I could see that the session had been a success. The mama goat was releasing her insecurities and the trepidation of being in her new environment. It was apparent that she was surrendering into her new role as a mama to a foster baby. No longer was she dodging and kicking the little baby goat away from her, and I noticed that her demeanor was softer and gentler, and her movements were not as stiff and jerky as they had been the previous day. The two were starting to create a visible bond.

As the hot days continued, I observed that the two were exploring one another on deeper levels and showing mutual interest in each other. They were building an emotional bond, and it all started with the HTA energy session. There was a different energetic halo around them, as though the rest of the world did not exist. They were slowly building a trusting relationship together.

During the last week of my stay, I enjoyed the experience as I watched the mama goat allow the foster baby to nurse from her. I strongly believe this is what integrating energy therapy is all about. It facilitates connection, healing, and "righting" the energetic pathways to flow, in turn creating a natural balance. Carol Gurney, a highly recognized Animal Communicator, who is also a teacher and one of my mentors, says that when we surrender judgment and closed mindedness and approach animals with willingness and humility, we are all able to understand what the animal is directly communicating. We enter a frequency, a space, similar to a radio station. "We are wired differently, animals and humans, but there is a place you can find this frequency, where there is a pure connection (Gurney 2001, xiv, 10-21)." She continues to speak about connecting to others from the heart space.

From there, we feel a "warm loving energy" which connects to our intuition. "It is the center, the core of who you are. Its language, what we call intuition, is a powerful form of communication... We trust reason and logic because they are more familiar and more valued by society and, therefore, more comfortable. However, reason and logic are limiting. Every time we ignore what our hearts are trying to tell us, we limit ourselves and our connections to others... Be aware of the depth and possibilities of the communication that can be reached between humans and animals (Gurney 2001, xiv, 10-21)."

I transcend into a full state of bliss every time I witness such an outcome, and I love being able to facilitate such meaningful changes as an energy practitioner.

## Chapter 6

# SHAVITRI "DANCING QUEEN" & CHANCHEL ~ FEMALE ELEPHANTS

*Human morality is unthinkable without empathy.*

~ Frans de Waal

## *Shavitri*

Shavitri is a 27-year-old female elephant who I have fondly nicknamed, "Dancing Queen." Her main Mahout's name is Mama, and she is very close friends with Chanchel, her 42-year-old stall mate. Even though these two are not related, they might as well be. They are very much attached to one another and always need to be in close physical contact with each other. Even during an HTA therapy session with one, the other joins in organically. These two share a trusted and deep emotional bond, in addition to a karmic connection.

I gave Shavitri the nickname Dancing Queen because she exhibits an excessive swaying behavioral pattern, constantly moving back and forth, placing weight on one front leg and then the other. I term it as "excessive" because this movement is much more pronounced in her than in any of the eight elephants at the elephant barn. After meeting and observing around 30 other elephants residing in the Elephant Village a few miles away, Shavitri definitely stands out.

It's not uncommon for elephants to move or sway. They generally move all of their body parts around. This is a common practice, and it has a purpose, keeping the blood and body fluids circulating through their impressively large bodies. They flap their ears, sway their tails, switch their weight from one foot to the other, and so on. They do not tend to stand still for very long periods of time. The rate at which they move, however, is different and varies according to their emotional state and mood.

Shavitri, however, stands out because she is almost in a constant swaying motion. She is one of the youngest elephants of the clan at the elephant barn. When I first spent a few weeks with this same elephant family in the fall of 2017, she was extremely restless and was not very affectionate with any human being other than her Mahout. Her demeanor was removed and distant, with a fierce trepidation of getting close to anyone. When approached, she would hold her head high, raised back and with a strong stiff posture, showing disinterest. Her body would sway with deliberate tenseness, a sense of vigor, and solid focus. She backed away from people rapidly if they ventured too close too soon, and she had no desire to come close to us. She gave us obvious signs that she was uneasy and needed her space. Her eyes were bright, wide open, and acutely aware of her surroundings, and her "state" would communicate quite clearly that she was in full control, especially when engaging with others.

Because of her obvious trepidation around people, the HTA energy work I did with her during that time was usually done from a distance of a few feet away. It wasn't until I worked with the other elephants, allowing her to receive the energy frequency from a distance, that she became more trusting toward me. As a canine behaviorist and trainer, I know from experience that it is imperative for each individual animal to decide when they choose to engage with us. I do not force myself on them, as this is not a respectful approach and is counter-productive to developing a trusting relationship.

In September 2018, however, she allowed me to come close to her and even accepted my direct touch. For the first time, she showed that she felt safe with me and realized I was her friend and not a threat. She instinctively knew that my purpose was solely to make the animals there feel good. My overall goal was to ground her and lessen her excessive and constant movements. Even though she was a year older than she had been during my last visit, her movements still had a frantic energetic pattern to them. In the mornings, she was more physically active than in the afternoons. Every morning, after her bathing rituals, she would get her exercise by walking the many miles to and from Amber Fort. In the afternoons, when the sun's rays were blazing and the air was hot and stifling, she was still in constant motion. This unusual behavior intrigued me, and I wanted to get to the source of what was attributing to this pattern. After spending my hours with her, my theory was that by swaying back and forth, she was soothing herself from discomfort and some form of anxiety.

From my observations, I concluded that she was spending most of her time in her brain space, and the lower part of her body's energy system was somehow blocked, keeping her from grounding fully. I had brought my tuning forks with me for sound therapy on this trip, and I was excited to try them with her. These tuning forks promote relaxation, which would help her release the unnecessary excess energy that was no longer serving her and was keeping her locked in an ungrounded state.

## HTA Energy Therapy Sessions

After my first assessment with her, my reading, indeed, confirmed my initial presumptions. Her lower Chakras were compromised, and energy was not able to flow from these vortexes and connect smoothly with the others to optimize a succinct and precise channel of energy flow through her whole body. Interestingly, during later energy sessions, the lower Chakra centers remained open, and some of the higher centers became blocked. During the course of my work with her, I could see that her excessive swaying behavior pattern was calming down and her energy system was recalibrating. Not only did she attain a more balanced temperament and demeanor, she also easily adapted to the HTA energy sessions.

As we worked together, I found that she was also extremely attracted to the inhalation techniques using EO's, and some had an immediate visible effect on her. I would place a couple of drops of a particular EO on my palm and rub the oil between my hands to ground and connect with the oil before introducing it to her. When I walked over to her, she would meet me with her long trunk and smother her nostrils in my hands, inhale deeply, and absorb the scent for several seconds. Her head would bobble back and forth in a soothing rhythm, and I could tell she was enjoying this part of the energy session. In response, she would show immediate signs of relaxation, sway less, and I could see that she was present, ready to receive and connect with the Divine energy.

## Observations

Using the tuning forks with her for sound therapy was an incredible experience. Her strong and powerful body would gently lean toward me. For long moments, she would stand still without even flapping her ears, listening and internally connecting to each and every tone, as I struck each fork in a particular sequential healing pattern. She and I became one during these magical sessions. There was a lightness that lifted both of us to a state of exponential joy and euphoria. During this time, she was able to release fully; and due to her heightened emotional state, she would actually tear up. Her long lashes, wet with tears, would glisten as the late afternoon rays of sun filtered into the barn and shone off her face.

Another technique that she was particularly drawn to is called Etheric Heartbeat. This technique uses the abundance of heart Chakra energy to surround the whole body and clear any energy blockages within the body. Many of the elephants, including Shavitri, bathed in the light of this technique and soaked up every minute of it.

I was able to conduct six full sessions with her during the course of the month, in addition to providing her lots of extra love and affectionate moments throughout my days at the elephant barn. During this time with her, her movements became less frantic and were gentler in nature, and it looked like she was smoothly gliding back and forth with more ease and grace. We had worked hard together, and noticeable and drastic changes had taken place in her overall demeanor.

## *Chanchel*

Chanchel is the tallest elephant in the female clan. When you meet her, she is quite intimidating due to her incredible grandeur and statuesque frame. Behaviorally and physically, she is well balanced and overall quite healthy. She is also very confident in herself and will let you know what she likes and does not like. It takes her some time to warm up to others, but given time, she came to trust me completely.

Because she didn't have any physical or behavioral problems, she did not need HTA energy therapy for any major issue or trauma. Nonetheless, I integrated it to keep her calibrated and balanced, promoting her overall daily health and well-being. In addition, I wanted her to benefit from the energy work and take note of the effects this work had on her.

During the energy sessions with Chanchel, her stall mate, Shavitri, also absorbed as much of the energy as she could. Especially when I used the EO's for inhalation, Shavitri would lean in, rotate her trunk toward us, and gracefully expand her trunk to sniff fully and experience the energy exchange of the oil. The film footage I captured during these sessions is phe-nomenal to watch. There was a heartfelt connection between these two friends, who would push each other away with their snouts to be the first one to be close and sniff my hands with their huge wide-open nostrils.

Chanchel literally would dismiss a certain oil that she did not like, yet she intensely loved three EO's out of the batch of seven I had brought with me. The three she adored were Sacred Frankincense, Frankincense, and Palo Santo. In addition, just as Shavitri had, she couldn't get enough of the sound therapy with the tuning forks. Her relaxation responses were crystal clear every time I used the tuning forks, and it was obvious that these forks were extremely beneficial for her. She would stand stock still and stretch out her mouth, as though in a big yawn. Holding this stretch for more than a minute, she would then close her mouth for a few seconds and re-open it up wide again, holding it that way for another minute. This stretching cycle continued throughout each HTA session.

Just like her sister, Shavitri, Chanchel's water ducts were triggered, and she would drool from her mouth and her eyes would be soaked with tears. This show of emotion was so fascinating that the Mahouts and volunteers would gather around and watch the deep state of relaxation that she was experiencing.

When Chanchel moved her feet, she would be extra careful not to step on me. She showed such care and connection with me, and being in her presence, I would feel the Divine blessings penetrate deeply through my core.

During one session, as I neared her heart Chakra for a certain touch technique, she clearly let me know that she did not like my physical touch on that area of her body. I naturally followed her gestures and moved my hand to another spot on her body, intentionally transferring energy to her heart Chakra from this new location. Since energy transfer takes place all over the body, the exact location of my hands did not matter. And noticeably during all of the assessments, her heart Chakra was never blocked and was always vibrantly flowing. During other sessions, if I came too close, she would give me a gentle nudge with her elbow, which was the size of my whole torso, letting me know I needed to find another spot to place my hand. Her non-verbal communication style often made me laugh. Her clear signals confirmed that even though we did not speak the same verbal language, we certainly knew how to communicate with each other.

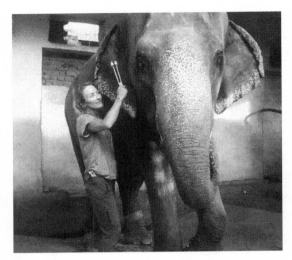

Shavitry

Chanchel

Chanchel

In other memorable moments with Chanchel and Shavitri, I observed their verbal communication, which consisted of deep, ground-shaking rumbling noises they would make out of pure joy. The middle space of their trunks would flutter while producing these intense guttural sounds. If one elephant started to "talk," the others would join in and communicate together with such bold, intense, and astounding harmony, their different pitches blending together perfectly, providing a complex musical sonata. I could feel their rumble chorus through my feet, and it would travel into my core and fill my whole being. Those were incredible times and magical summer afternoons at the barn ... and not just for me—everyone at the barn would stop and relish in these joyous occasions.

"Of the twenty-six documented vocalizations made by adult elephants, nineteen are made only by females, three are made by adults of both sexes, and only four are made exclusively by males. An additional six calls are made only by subadults. Of the twenty-two calls exclusive to females, nine are calls typically given in chorus with other family members, while thirteen are usually made by an elephant calling on its own. Females spend most of their time visiting and revisiting the experiences of communalism and only occasionally express their individual imperatives (Payne 1998, 97)."

At other times during the energy sessions, Shavitri and Chanchel would relax and release their trunks in unison, fully resting them on the ground in a half a circle. They are so in sync that there is no doubt that these two sisters are karmically aligned and soul mates. It is very evident that they are two parts of one whole.

## At the End of Each Day

At the end of each long and fulfilling day, having conducted about three sessions daily, I would download all the film videos captured with my camcorder that day, organize and back up the footage, and prepare it for further editing when back in the USA. This nightly process was my way to ground and connect even more with the elephants. With incredible clarity, I could see on screen the energy transfer, healing, and bonding taking place between the elephant, the Divine, and myself. I still treasure those evenings, thinking about all the amazing footage I watched and was able to capture. Now back in California, I am putting video clips together in powerful ways, voicing and promoting this magnificent work and sharing it with many others across the globe.

## Chapter 7

# BHITILY ~ Sister Horse to Rashma

*True love is when you touch someone with your spirit, and in
return they touch your soul with their heart.*

~ Anonymous

### First Day Back in Jaipur, India

Upon my arrival in Jaipur in September 2018, the monsoon
rains were still raging, and the streets and dirt paths were
wet and muddy. My heart was beating full of excitement
and anticipation, yet I was nervous to be back in Jaipur, because
I would be conducting a more rigorous research study and using
my camera equipment for the very first time.

This would be my first solo research study in the field of HTA
with elephants. Despite the anxiety, I could sense that this was the
start of a whole new chapter in my life, and with guidance from
the animals and the Divine, I felt a peace deep down that it would
all work out. I was also excited, because on Monday morning, I
would see my female elephant family again and meet any new
characters hanging in and around the barn.

I could barely sleep that first night, and on Monday at 12:30 p.m., I was picked up by Totoram, my friend and tuk-tuk driver. We hugged in a reunited embrace before we proceeded to drive through the colorful and loud streets of Jaipur. From there, we turned onto a semi-highway, and then after about 15 minutes, turned left onto an unmarked dirt road. Here, the mud path was bumpy and muddy, and we hobbled along through a few tiny towns on our way to the elephant barn. On a portion of this winding road, there was a straightaway, a nicely paved road where Totoram was always able to increase the speed of the tuk-tuk. This was one of my favorite parts of the journey as I felt the warm air gliding over my face and was able to treasure the sights along the drive. Some of them were incredible—on my left, there was an old, breathtaking pink Hindu temple overgrown with weeds and grasses. I always wonder about its history.

My heart continued to race during the ride, and I smiled brightly with disbelief that I was actually back—back to a place I had come to love and had become a part of. To be more accurate, this place had become a part of me. We passed through an arched medieval gate with massive rundown and rustic wooden doors that were decorated with bold iron designs. Overgrown bushes with bright pink flowers dressed this dilapidated structure. It was like a scene from an ancient movie. As we drove on, townspeople, often barefoot and dressed in colorful clothing, would walk alongside the road for miles, carrying big water jugs or huge stacks of grasses, hay, sticks, or yarn on their heads. Their children, with huge deep brown eyes and big smiles, impeccably dressed in their bright white button-down shirts and plaid ties that were part of their school uniforms, would stare and wave me on excitedly. The warmth of their outspoken social nature spoke volumes. This welcome made me feel like I was accepted, a part of their day-to-day life and their community.

The air felt warm, but still wet from the rainy days that had passed, and we finally arrived at the elephant barn. Upon disembarking, I saw a thin white horse in the distance, eating what looked like minimal greenish brown weeds. As I stepped down from the tuk-tuk, I saw all the other animals that also resided at the barn, the cows, dogs, and goats, and felt a rush of excitement. It is quite common in India for many different species to coexist with the human population. For example, in the town and even in larger cities, monkeys roam the streets and climb the houses to watch the world go by from the rooftops.

## HTA Sessions Week One

Once the Mahouts, elephants, and I exchanged our welcome greetings, I embarked upon setting up my camera equipment and getting to work. Even though I could not wait to start structuring my schedule and connect with the elephants, I was first drawn to meet the horse that I had encountered upon arrival. My Mahout friend, Atchou, is her guardian, and he let me know that her name was Bhitily and she was 10 years old. We went outside, and he called her over. As she walked toward us, I could see that she looked much too thin. A deep rush of fear flooded my body, and I tensed up. The prior year, I had worked with Rashma, Bhitily's sister, who had lost her life that same year, most likely due to a lack of nourishment and insufficient care. I couldn't believe that Bhitily was also this thin and undernourished. I held back my tears and tried to refocus my emotions so I could consider the actions I could directly take to provide a positive impact and create change within her. I asked Atchou if he would allow me to work with Bhitily on a regular basis, and he gratefully obliged, saying he was worried about his horse. My mind raced as I thought about what I had to do. I needed to devise a holistic and inclusive "back to health" plan, where Bhitily would gain more strength and weight, in addition to regaining her vitality and help her become less skittish and more grounded.

First and foremost, I would observe and assess her fully to design a rough treatment plan for the upcoming month and start HTA sessions right away.

Atchou tied her to the gate attached to the elephant barn, and I gently made my way over to her, while remaining at a safe distance so she would not feel threatened by me. I had plenty of room to roam around her. I gave her space and time to get used to my presence while I observed her up close. Then I increased my distance, giving her some room before she decided to move away from me, and then I slowly drew near her again, all the while observing and taking notes. This technique allows the animal to get used to my presence without forcing myself upon them. She observed me with extreme caution, and I noticed her huge eyes following my every move. I could tell that I needed to be gentle and take my time in getting in closer proximity to her physical body. During my observations, I noticed several undoubtedly infected wounds on the bones protruding on her thin body. Flies were swarming the wounds and her face, making her edgy and irritated. She was constantly swaying her tail back and forth, nudging at them with her snout, and her body was so irritated by the flies that she was shivering. This poor girl needed compassion and care on many levels. Her overall demeanor and behavior patterns were static and edgy, indicating that she was in complete discomfort. I was going to conduct many HTA sessions with her and bring her extra food daily to increase her strength and inner fight toward healing, just as I had with Rashma the year prior.

During my first assessment of Bhitily, my observations were as follows: 1) Bhitily was extremely thin, and her hip and joint bones along her spine and ventral part of her body were protruding dangerously; 2) She had open wounds on her bony frame, along her spine and hip bones, and flies were feasting on her fleshy wounds; 3) Her spirit was broken, and her energy system was not in one coherent vortex, but instead presented a shattered and scattered nature; and 4) Her behavior was skittish, unsure, and distrusting.

At the hotel later that evening, I wrote down the list of healing practices I would need to integrate with her and a list of all the items I needed to purchase to rapidly enhance her healing. I planned to: 1) buy or make an organic fly spray to keep the flies away and prevent their feeding frenzy on her infected wounds; 2) heal the infected wounds topically using saline solution, clean dressings, and Copaiba EO, and make sure the infection inside her body would stop simultaneously; 3) feed her more wholesome foods, starting gradually and increasing at a pace that her digestive system would be able to handle, thereby increasing her strength and building more body mass. I was also going to purchase mineral and vitamin supplements to add to her diet; 4) build a trusting relationship with her that would calm her, ground her, and facilitate my energy work with her on a regular basis; 5) re-energize her spirit and sense of belonging on this earth so she could regain strength and heal through purpose, connecting to her intuition and animal instinct, which now seemed to be energetically blocked; and 6) mentor and convince Atchou to take better care of her. As a foreigner, I knew I needed to remain objective and without judgement. I wanted to stay neutral and have a positive effect on Bhitily. The only way I was going to be able to make some positive changes would be in conjunction and collaboration with Atchou. I was very confident that by the end of the month, Atchou and I would have shared and learned from one another and would be in agreement regarding the best care for Bhitily and the other animals residing at the elephant barn.

Bhitily and I embarked on our journey together, healing and bonding as I brought my best self forward every day. Together, we completely turned around her physical, mental, and spiritual state. I was able to conduct six full HTA sessions with her, including intermittent shorter daily visits, and adding in one or two HTA techniques, such as Vibrational Grooming and Magnetic Clearing, in addition to observing, feeding, and caring for her.

Every afternoon, while in transport to the barn, Totoram and I stopped at several fruit stands, bartered for the best deal, and bought at least 2kg of bananas for Bhitily and the elephants. Once a week, we stopped at an all-purpose big windowed store-front to buy a bag of hard sugary donut holes for them, as well. I always giggled at the way the animals at the elephant barn liter-ally drooled when just looking at and smelling these donut holes. I had learned the prior year that Rashma, Bhitily's sister horse, loved bananas. I realized then that feeding bananas was an easy way to add nutrition to their regular diet of grasses. Like Rashma, Bhitily also enjoyed eating bananas. I was happy to bring several kilos daily for all the animals residing there.

## Supplies for Bhitily

In India, it is quite difficult to get around town to buy regular sup-plies and run errands. I tried and learned my lesson the hard way. On my first Saturday, after my first week of work at the elephant barn, I traveled with an Indian friend, going all over Jaipur in search of the supplies I needed. One product I wanted to pur-chase was horse feed. My only option was to purchase a year's supply right there on the spot. Not only was it very expensive, but I did not see the benefit of purchasing in bulk and letting this feed just sit outdoors where it would be unprotected. The logistics were not ideal.

Another complete failure was my ability to purchase the ingredients needed to make homemade anti-fly spray. We failed miserably, and I was exhausted by the end of the day, having not made any purchases. Back at the hotel that evening, I tried to ease my frustration. Hence, I went online to see if I could order products, instead of trying to find them locally. Through Amazon.in, which I didn't know existed until then, I was able to buy some of the items I wanted and have them delivered to my hotel. I couldn't believe my luck. My spirits lifted tremendously, as I was finally getting concrete results from my efforts to aid the animals.

Later, I found out this was not completely true, but I was able to order and receive at least 50 percent of the provisions that I needed for caregiving. The reason for this 50 percent margin was always for some unknown reason; sometimes the packages were delivered, and other times, they never showed up without any valid explanation. My rational understanding regarding this matter is that India is wonderfully unpredictable and one learns not to get frustrated and to have a lot of patience. In the end, I did receive a 6-month supply of vitamins and mineral supplements, as well as extra Copaiba for Bhitily. I was completely thrilled that at least these supplies had made it to my doorstep and that I would be able to continue to provide aid after my departure.

## HTA Sessions Week Two, Three, and Four

Back at the elephant barn, I was working with Bhitily and her healing process, just as I had with Rashma the prior year. I addressed the wounds with saline solution and Copaiba EO, which due to its antifungal, antiviral, and antibacterial properties, is effective in fighting internal and external infections. This amazing oil can be placed directly onto wounds, cuts, and scrapes to promote healing.

During the first week, I was still not able to get close to Bhitily and I had to have Atchou tie her to the gate and distract her in order to directly dab Copaiba on, or at least close to, the wounds. This proved a little challenging at first, but working together, we got the hang of it.

After more time spent just being in her presence, doing some light Vibrational Grooming, and speaking with her, she did allow me to place the Copaiba directly on her wounds without Atchou's help. Using slow movements and gentle energy, I applied the Copaiba when she allowed me close for even a few seconds. Later, she even allowed me to rub the oil directly around her wounds. Not only was she becoming more accepting of me and my treatments, but Bhitily was becoming vibrant again, as well. My fears about her subsided as I observed her rapid initial transformation in only a week's time. I had mixed and integrated all the tools available to me in a holistic healing approach and could see that the effects were positive thus far. I do not value one technique or concept over another. I believe that on an integrative whole, we can accomplish much more than if we choose to only use one technique or one medical practice or set of techniques at a given time. Using the tools we have at our disposal is empowering and allows us a greater ability to provide the animal with optimal care.

As the heat and sun intensified in the afternoons and the days flowed from one to the next, I settled into my work and routine. I found myself observing and spending quality time in Bhitily's presence each afternoon, as I made the rounds with every animal there. When I did not have her scheduled for a full HTA session, I conducted certain solo techniques that would benefit her and advance her healing process, in addition to deepening a more trustworthy animal-human bond.

Despite this quiet, consistent work rhythm, I never knew what to expect from one day to the next. That's how India is; it is a culture of color, chaos, and noise, except at the elephant barn, where it was quiet and peaceful.

I was better prepared the second time I visited India. I knew what to expect and was drawn to this chaotic, yet simple, life, which is very different from my Western lifestyle. Spending time with the animals on a regular daily basis gave me a way to ground and remain balanced, so I could be most present and effective in my HTA sessions with them.

One technique that proved to be very important for Bhitily is Vibrational Grooming. This technique requires the practitioner to lightly touch the body, stroking the whole body, hand over hand. In the beginning, while we were getting more acquainted with one another, there were moments when she was accepting of my very light touch and others when she would intentionally step or canter away from me. I had brought some "grooming" gloves with me that I purchased after seeing them in an ad on Instagram. These gloves have little nubbles on them to stimulate the skin while grooming in such a way that many animals love. I used them with her and also used them to bathe and clean the elephants. These gloves have additional benefits in that they help to increase the blood flow to the surface of the skin, thereby energizing the cell structures.

Another technique that attracted Bhitily was sound therapy using the tuning forks. Her stiff and rigid body would soften and move into relaxation mode during these sessions. The film footage I shot during the sessions showed an incredible transition in her body structure and demeanor, triggering her physiological relaxation response. When I placed a tuning fork on her spine with the vibration penetrating into her body, she would lift her head with purpose and become alert and present. Then she would allow the vibrations to penetrate her body and mellowed into and blended with these sensations. When I placed the tuning forks a few inches from her ears, she would lower her head and with heavy eyelids, her eyes would become soft, her lips would part, and she would start chewing, drooling, grinding her teeth, yawning, and letting out her own unique sounds of energy release.

She started to allow my touch regularly and with ease. I was able to directly apply hand contact all over her body for longer periods of time, gradually increasing the pressure, even around her healing wounds. Thankfully, as her wounds healed, the flies were also less attracted to her. She became less aware of them and less irritated with the few that were still buzzing around her, as is normal at the elephant barn. Luckily, her wounds were no longer infected and were healing rather well.

It was no wonder she started to trust my presence around her; simply put, she was starting to feel better. She wanted more daily attention and would come looking for me, even when I was working with the other animals inside the elephant barn. She always settled into our sessions quickly and accepted my touch with ease and grace. In front of my eyes, I could see her healing process on all levels.

As the days moved forward, she and I became so in tune with one another that everyone at the elephant barn started to call me Bhitily's mama. By the end of the second week, she was untethered during all the remaining HTA sessions, letting me touch her and stand close without moving away from me for any reason during a full-hour treatment session.

I never know what to expect when I do this work. Each session is different, and each animal shows up with their own unique and true personality and character. I enjoy reading their body language and listening to their unspoken communication. This allows me to be fully present for their needs and benefit, and it takes away any human ego judgement or expectations that I might have. In addition, it allows a trustworthy relationship to form more rapidly. I am not telling them what I want them to do; I listen to what they need. This concept is the only one that works for me and the animals I work with. I believe it shows the utmost respect to other species.

Bhitily

Bhitily

## Transformations

Bhitily's transformation throughout the month was incredibly successful in so many ways. My complete holistic approach and treatment plan had benefited her. Her physical body had filled out, her rib cage was less noticeable, and her overall demeanor was less scattered and skittish, and more centered and grounded. She showed a strength in character, and her true instinctual soul was present once again, boldly shining through. She was more aligned, representing herself more fully and wholly as a balanced being. Her wounds were no longer infected, and even though her skin was still light pink, I knew her hair would start to grow back in time. I also noticed that she was more social with the human beings around her. She showed interest in others and would peek into the barn when I was working with the elephants and peacefully observe. She was a smart girl and would hang around any HTA energy sessions I was conducting with other animals, reaping the benefits hands on or via transference through another animal.

The kids who resided around the elephant barn had been watching and observing my care. They helped me feed Bhitily, and I hoped they would continue to take good care of her. The Mahouts also played an active role in helping me provide for the animals at the elephant barn, and they had spent hours observing my work. While drinking chai during afternoon breaks, we would discuss and comment upon the positive impact of the healing work and the noticeable changes that were taking place with the animals. We were all working together, which was beneficial for me and for the animals. I wanted the Mahouts to not only experience the work, but to see the value in it, as well. Knowing my work would hold more value if it continued after I was gone, I taught them some simple energy transfer techniques they could easily implement and continue to use in my absence.

During my last few days at the elephant barn, "Mama," Shavitri's Mahout, would take my video camera and film me as I conducted my sessions. As he filmed, everyone laughed with delight as they observed Bhitily following me around and being my shadow. She had come an incredibly long way in the past month. We had established an everlasting bond and relationship, and I will remember those days we spent together fondly. I was already looking forward to seeing her on my next visit to Jaipur, knowing it would be a delight.

I left the Mahouts with a six-month supply of nutritional vitamins and supplements that easily blended into Bhitily's daily water. I also left them with several vials of Copaiba to use with any of the animals (and for themselves, as well) to fight off bacteria and infections, strengthen the immune system, and keep everyone healthy.

## Reflections

Reflecting back on the time spent, I feel complete bliss. Even though we don't speak the same verbal language, as we connect with animals with our senses and intuition, we come to realize that there is so much depth beyond verbal communication. Through close contact with our animal kingdom, we actually learn how to reconnect to our own powerful intuition and instinct. Robert Greene takes this one step further in his book, *Mastery*. He conceptualizes for the reader that intuition and human rationalization do not have to be mutually exclusive. On the contrary, they can seamlessly coexist. When we spend time and focus on our devotion, increasing our knowledge base in our subject matter and further developing our analytical skills, our "intuition springs from a high rational focus (Greene 2012, Chapter 2)." Through our experience in intuition and insights, we will reach elevated degrees of reflection and reasoning, just like Jane Goodall, who spent years of focused observation, studying, and learning from the chimpanzees in their environment, not in a laboratory setting.

For me, it is completely rational and practical to integrate animal energy therapy practices with all species across the globe. I am more than rewarded from spending my time during my travels, and in my day-to-day life, connecting with animals in this manner. I see and feel the constant changes that occur daily when doing this work. Spending several months in Jaipur solely doing HTA therapy was yet another strong affirmation for me. This work is amazing, and I feel blessed to be an instrument to provide this care to as many as I can. Optimal health and well-being of the body, inside and out, needs to be observed as a whole unit.

Utilizing all the tools we have at our disposal and making rational sound judgements pays off exponentially. We have Western medicine and Eastern therapies, which rationally should be used together in every way to provide the best health and longevity of any animal on this planet. We need to be reminded that we are all stewards of living beings, plants, and creatures on this planet. As stewards, we need to keep taking action to care for the planet and all the species that reside on it, as this is a gift to mankind.

With a more open-minded approach, we learn and grow from selflessly serving others. By listening, being present, and giving, we will heal through our animals as they teach us and show us the way. As guardians, lovers, and advocates, we are able to collectively raise the frequency of the planet, and our children will be able to continue to cherish and experience the beauty of nature and wildlife on this planet.

## Six Months Later

In May 2019, I received a text from Sukret, the co-founder of Volunteer with India, letting me know that Bhitily had a newborn. I also received several photos of the two of them together. Tears of pure joy streamed down my face. Her body to this day remains healthy, and I was relieved that she was able to carry a healthy and strong foal. I can barely wait to meet her new offspring and am already starting to plan my next trip to Jaipur to visit the Indian human and animal family I was lucky to bond with. Of course, while I'm there, I intend to celebrate the birth of Bhitily's foal, Aahana, which means inner light, immortal one, and born during sunrise.

## Chapter 8

# CHAMPA 2 & RUPA ~ FEMALE ELEPHANTS

*Love and compassion are necessities, not luxuries. Without*
*them, humanity cannot survive.*

~ Dalai Lama

## Champa 2

I fell in love with two female elephants named Champa 2 and Rupa. Both are strong, emotionally sensitive, beautiful, and centered ladies. Champa 2 is the smaller of the two, and she is 20 years old. Munna is her Mahout. Rupa is 40 years old, and Kahn is her Mahout. The dichotomy of strength and power next to sensitivity and sweetness that both of these elephants displayed overwhelmed me. Their direct and unwavering gaze into my eyes and soul penetrated deep within me. Champa 2 has glorious blue eyes that are mixed with some hazel rays. Due to her rather small size and stature, she would lower her head to look straight into my eyes and place her brow in close proximity to mine. We spent many moments staring at one another on a regular basis, and I enjoyed seeing her watch me with affection and affirmation. Just as it had been evident with the other elephants, it was clear that Champa 2 and I also did not need a similar spoken language; our communication was dynamic and present. In the beginning, because her stare would be unwavering and direct, I felt I needed to fill the space with words. However, as soon as I worked through my own trepidations, I learned that speaking was unnecessary, and I became more in tune with silent, intrinsic, and instinctual ways of communication.

Before conducting my HTA sessions with Champa 2, I mentally asked her if she was willing to be a participant. Through her strong and solid gaze into my eyes, she communicated to me her desire for this energy exchange and deep connection. True to her unspoken word, Champa 2 welcomed any and all of my techniques. She allowed my touch on her body and was attracted to aroma and sound therapy. As soon as she and I entered her barn, there was a closeness in our energy exchange and connection. We understood one another and looked forward to getting to know each other more every day. We grew so close that it became difficult to say goodnight to Champa 2 after our visits. Her magical and magnificent being put a spell on me, and time stood still when I was with her. I would place my nose on her trunk regularly and take in the incredible smell of her skin. All of the elephants had this particular scent, and I loved the sweet smell of their skin.

## Champa 2 & Rupa

Champa 2 and Rupa adored being touched, and both would snuggle in close to me when I conducted HTA sessions with them. They literally swayed, leaning on one leg directly into or close to my body. I was able to conduct six full HTA sessions with both of them during my one-month stay, in addition to spending stolen moments to observe and bond with them. That's how much these incredible elephants mesmerized me. Both of them immediately accepted me and took me in as their friend and energy therapy practitioner. They opened their energy fields into mine, trusting me completely and wholeheartedly.

As is common and as I experience daily in this small female clan, the bond between these elephant friends is very strong. They verbally communicate with one another often, and the loving connection they share is clearly noticeable. What is unique is that Rupa and Champa 2 did not have a loving bond. They were actually housed in adjoining barns and have gotten into physical fights with each other.

Common factors that exist in the same female clan led by a matriarch are: 1) they all have strong female bonds with each other and friendships, 2) there is constant socialization between them, and 3) they are family members for life. Despite their incredibly gentle disposition and loving nature toward others, Champa 2 and Rupa would fight each other if given the chance and for that reason are housed in different barns.

Both Rupa and Champa 2 are physically healthy elephants, and  both have had some temperamental issues when interacting with one another and with humans. Rupa, who is almost twice the age of Champa 2, has literally "grown up" and moved past her anxieties around humans and various caregivers, but Champa 2 still displays anger and distrust. Munna is the only Mahout who can work with her. Munna lives by her side day in and day out and is very close to her. He and his family sleep in one bedroom adjacent to her barn. I am lucky that the two of us get along easily and that our energy together is solid and steadfast. Her trunk is so strong that she can easily throw me across the room if she didn't favor me in any way, yet she has never shown any sign of discomfort or dislike toward me.

Munna and I would regularly speak about elephant body language and communication, which provided me with the knowledge I needed to make sure I always acted with respect toward them. I soaked in all the details of elephant verbal and nonverbal communication, wanting to know as much as possible. From what I learned, I was delighted to feel and know that all the elephants at the barn were happy to see me, including Champa 2.

One time when I was conducting an Energy Session with Champa 2, I showed her my cupped hands for a few seconds before following through with my intention to place them on her heart space, or heart Chakra. She lifted her trunk willingly and allowed me to step in, closely fitting my body underneath her chin, and allowing both my hands to rest on her heart Chakra for several minutes, without either of us moving. Thereafter, I gently stepped back, opening up my arms, transferring the energy flow from her heart and directing this glow to radiate over her entire body for about 20 minutes. Closing the technique, I brought my hands together again, one on top of the other, and walked back to her. She again lifted her trunk, allowing me to connect back to her heart space.

We moved with unspoken synchronicity through many similar techniques, flowing together in unison. Due to this incredible communication together, the energy sessions were intense and very powerful. It was as though she was able to read me, and I was able to also read her. She understood the steps of the technique and welcoming the exchange.

Rupa was the only other elephant there that I had similar experiences with. She also read my signals and corresponded in the exact same way with her behavior in order to facilitate the components of each technique and optimize the benefits of the energy sessions.

Rupa wore her heart on her sleeve and boldly showed her love, leaning in, cuddling up close, closing her eyes, and refraining from movement. The two of us, in complete silence and lost in each other's essence, shared moments of pure energy exchange. And just like Champa 2, no matter where I placed my hands on her body, she was completely receptive to my touch.

Rupa is also quite the jokester and would sniff in the flour that was used to prepare the chapati (the bread baked every couple of days for the elephants), The Mahouts made this dough on a stone platform right in front of her stall space. Trying to avoid being noticed, she would slowly move her big nostrils over to the flour and take in a big sniff, holding it for several seconds as she lifted her trunk, and like a fountain, she snorted it all out, creating a cloud of floating flour that dusted all of us nearby. Her playful nature would continue with her inquisitive behavior, and she'd lift her trunk toward the camera as I was filming her. She swayed her trunk over in a gentle manner and, intrigued, placed her nostrils as close as she could. I still laugh when I look at the video footage of her as she interrupted the energy session with her playful nature. Animals do not play when they are not in a state of ease and happiness. I experienced her happy state all the time, which spoke volumes about how good we felt and how happy we both were in each other's presence.

Because Champa 2 and Rupa were individually so friendly and welcoming toward me, I was surprised that Champa 2 and Rupa did not get along with one another. Their independent personalities were so full of love and life, but when I first arrived, they were incompatible and couldn't be together. After my first visit in 2017, and after two weeks of HTA energy work, they were able to be on opposite sides of the same barn together as comrades. One afternoon, the Mahouts wanted to surprise me with the good news

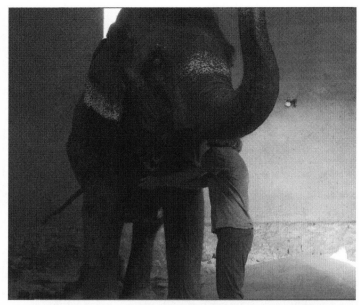

Champa 2

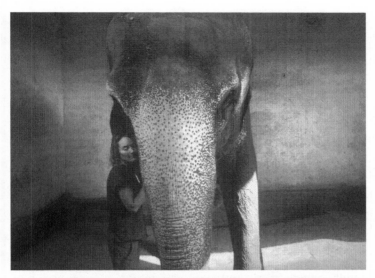

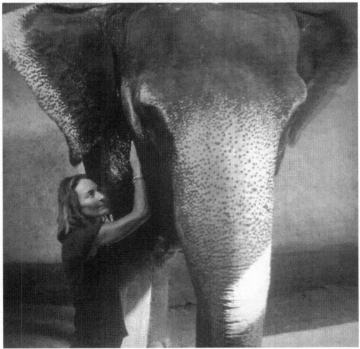

Rupa

that both were able to be in closer quarters together. This really was great progress. Perhaps they were respecting one another and each other's space. They were not being quite friendly yet and perhaps were just ignoring and tolerating one another, but at least they were hanging out in the same barn together without showing any form of anxiety. I attribute the HTA energy therapy to this success.

## Observations

After working with these two elephants, there were two main observations. The first was that they became more accepting of one another, and the second was my ability to validate which specific techniques had the most profound impact. Learning this was extremely beneficial for me. After my first visit to India the previous year, I chose the specific techniques to use for my explorative research study the following year for maximum benefit and consistency. All HTA techniques are valid and useful in many different situations, but for research purposes, I wanted to keep my variables to a minimum. As my research will continue with more animal species on different continents, I will have a better sense of how to use fewer techniques, being concise in my variables for publication purposes and obtaining research grants.

## Insights from Others

Paul Barton, a British classical pianist, has lived in Thailand for the past 22 years and has been playing classical piano and flute to elephants who reside at the Elephant World Sanctuary. These elephants were abused, sick, and retired; they had been rescued from tourism labor, the logging industry, and the streets. In his films, he shows the impact that music has with these magnificent elephants. In an article on the Coconuts Bangkok website, he states that Plara, the first elephant he played for, enjoyed the slow classical music he played for him. "He curled his trunk and held the tip trembling in his mouth until the music was over (Sakaowan 2018)." Others would stop eating their grasses and listen to the magic of the sounds he was creating. He continues to state that "almost all elephants react to music in a visible way. There's a sudden movement when the music begins. The elephants are free to walk about around the piano; they are not chained or tethered in any way. If they didn't like the music, then they could simply wander off. Some elephants get very close to the piano of their own accord. They might drape their trunk over the piano even. Some hold their trunks in their mouth when listening, some start to sway with the rhythm of the music (Sakaowan 2018)."

Temple Grandin connects language and music with animals, as well. She mentions in her book, *Animals in Translation,* that evidence suggests that "music developed in animals long before humans evolved." This evidence comes from a study of animal music by Patricia Gray, who is a pianist of the National Music Arts program, and five biological scientists that was published in the prestigious journal, *Science.* "The fact that whale and human music have so much in common, even though our evolutionary paths have not intersected for 60 million years, suggests that music may predate humans – that rather than being the inventors of music, we are latecomers to the musical scene (Grandin 2005, 278)."

Using the tuning forks with both Champa 2 and Rupa, I noticed similar reactions to those Paul Barton mentions. Both of them, as well as Champa 1, would lay their trunks on the ground in a little curl, resting and relaxing their trunks in direct response to the sounds they were internalizing. All of the elephants would sway into me, listening and connecting to the sounds. They would lean in close, stop flapping their ears, and just take in the healing sounds.

This work allows us to connect with all the senses, making us able to generalize the reactions shown by certain animal species when these various forms of energy therapy affect them and trigger them into a state of relaxation. Being in this state results in the rebalancing and recalibrating of the body's energy channels so healing can take place.

Amy Snow and Nancy Zidonis, two very insightful teachers who are also my direct mentors, founded Tallgrass Animal Acupressure Institute. They have written books and taught many students the step-by-step guide to animal acupressure. In their book, *The Well-Connected Dog: A Guide to Canine Acupressure*, Dr. Michael W. Fox, a veterinarian and author states, "There are other tools of alternative veterinary medicine and companion animal health care maintenance that accord with the bioethical principles of this new (and ancient) approach to improving animals' health and well-being. Much improvement is needed and will be achieved when certain bioethical principles that constitute the rights of all animals under our dominion become part of the heart and moral fabric of society. These principles are: right breeding (to avoid harmful diseases of heredity origin); right socialization and rearing; right handling and understanding; right environment and nutrition. These are animals' basic rights and are our cardinal duties as their caretakers and custodians (1999, Foreword)." He continues to note that by harming just one or a whole species, ecological life and community, we upset the balance of universal energy and the Chi matrix. By doing so, we harm ourselves. "This is the karmic, anthropogenic component of most diseased conditions, from cancer to catatonia, that afflict our species and many other species today." We need to "try to do whatever healing and prevention and alleviation of others' suffering that we may in the course of trying to stay in balance ourselves... St. Francis of Assisi, among a few other saints and sages, manifested this divine power through this mutual affinity with fellow creatures: Holy communion indeed (Snow and Zidonis 1999, Foreword)."

Through my experiences with all the animals on my path, the effects of the energy therapies are evident and need to be integrated with all species. As Martha William states clearly in her book, *Learning Their Language* (2003), we have to be in partnership together, with care and mutual respect for healing to take place. Of course, we need to learn from and listen to them as we enhance our animal–human connection and expand our human capabilities in our lives, choosing purposeful living that is positively connected with the animal kingdom and nature.

## Chapter 9

# JAI "INDIAN PRINCESS"
## ~ FEMALE DOG

*Although other animals may be different from us, this does not make them LESS than us.*

~ Marc Bekoff

Near the end of my trip, I decided to rescue and adopt a street dog that I named Jai, which is short for Jaipur, the city that I was living in at that time. In Sanskrit, this name means "victorious." Of the many street dogs that I encountered in India, this one-year-old little black pup with a white-tipped tail and white feet captured my heart.

I met Jai the first day of my arrival in Jaipur. A friend picked me up from the airport, and when I climbed out of his car, this little dog came running up to me, greeting me with a bow, followed by a quick dainty jump up to say hello. She was actually grinning at me, and I saw right away that her hazel eyes were bright, clear, and happy. I was a little startled at first by her as

I had not experienced such friendliness with street dogs in the past. Normally, they don't venture over with such delight and such a warm greeting.

From that first day, our bond and relationship developed. There was a connection between us that grew over the days that I would see her, even though some moments were only fleeting. I saw Jai only in the early mornings and in the later part of the evenings, when the heat of the day had slightly subsided and the sun was not as blazing hot and intense. Twice a week, on Tuesdays and Fridays, I would rise around 4:45 a.m. to get ready for my journey to the elephant barn so I could wake the elephants and assist the Mahouts with their daily morning ritual, getting the elephants bathed and dressed for their morning excursion to Amber Fort. I would wait in the morning darkness for Totoram in his tuk-tuk to pick me up and take me to the elephant barn. In the cool morning air, I would sit on the curb next to the deserted street, outside the gate of the hotel. This street was the main thoroughfare, and while it was quiet during that time, it would come to life in only a few hours. There was a little side road that ran perpendicular to this main street, where fruit vendors would set up and congregate daily. A half a block or so from these vendors was a little alcove leading to the entrance to a local bank. This is where Jai would sleep during the night.

Since I provided the only stir in the early mornings, Jai would venture over with a light, sleepy gait for a morning greeting and lie down next to me, keeping me company until Totoram arrived. She would snuggle next to my legs and let out a sigh of relaxation as I stroked her body. There was no doubt that she was quite a special little girl. She also looked healthy. Even though her coat was dull and lacked luster, her eyes were clear, her teeth were white, and she had no foul odors emanating from her body. As our special moments passed, I started to realize that she and I shared a karmic and special bond.

My residence for the month was a corner room with windows looking out to the little side road where Jai would wander and hang out underneath the fruit stalls and vendors to escape the sun. Because I was in such close proximity to the places Jai ventured, I started to look out for her in the evenings. When I saw her, I would walk out onto the overpopulated sidewalk and along the street, greeting her and spending time with her in the midst of the chaotic crowds and activities. During this time, I was able to observe her and share short moments bonding with her on a daily basis. Our connection continued to grow throughout these early morning and late evening hours.

The weekend before I left India to return to California, I decided that our connection was so strong that I wanted to adopt her and bring her with me to the United States. Logistically, this was a tremendously difficult and tedious undertaking. I had to take care of a lot of details with veterinarians and government visits, gathering the necessary paperwork in order to bring Jai home with me. India is not an easy country to live in, nor is it an easy country to adopt a street dog from. I kept running into roadblocks that I somehow overcame with pure persistence and boldness, becoming quite demanding in the process. I was so extremely passionate and felt so aligned about this adoption mission that I was able to tap into my supernatural power utilizing Divine guidance, and, against all odds, I connected with a few instrumental people who not only spoke English but who also understood my quest.

## Rescue Mission and Energy Therapy

One Friday evening around 8 p.m., I executed my rescue plan for Jai. It was a task and one that was well planned. I plucked her from the streets, took her to a vet clinic for a preliminary check-up, and then dropped her off at a boarding facility. I had made arrangements with my friend, Govind, to have his friend, Roop, escort Jai and me in his tuk-tuk around this big and busy city of Jaipur and fulfill my plan. The plan could work, but first and foremost, everything depended on me being able to find Jai and contain her for transport. Without Jai, there was no mission. There was a lot of driving to do throughout the big city, and we were in for a long evening.

All the little details of this major plan kept my mind racing, and even though I was completely anxious about how the plan would actually transpire, I also had an incredible amount of adrenaline. I cut a leash and another long piece of fabric out of a pair of cotton pants, which I turned into a harness to use for her capture and first transportation ride. I also took a picture of the address of the vet office, since I would not be able to communicate with Roop, due to neither of us being able to speak each other's language.

I spotted Jai from my window, and the adventure started. This plan in the making needed to be fluid, but it was nighttime, and I only had so many hours to make it work. If I failed, Jai would land right back on the street. I ran down the stairs of the hotel, through the lobby, and out into the hot night air and the disorderly, crowded street, and Roop ran over for an initial greeting and helped me search for Jai. I spotted Jai about a block away, where she was running and playing with some other dogs. We ran over to her, and as soon as she heard my voice, she came running to me. I was shaking with nerves, and it took me a little while to get my homemade harness and leash tied around her. It was dark and Jai is black, and my hands were shaking with trepidation and excitement. However, I somehow managed to successfully take control. I didn't want my attempts to capture her scare her and cause her to run off into the busy street. Once she was tethered, we coaxed her to the tuk-tuk, and she actually came quite willingly. She sat between my legs, and I think that's when I took a breath for the first time, as a first wave of exhilaration and relief flooded over me. This was only step one; there would be many more to come, but at least she was safe in my arms. As the night grew cooler, I stared out at the streets that were new to me as we drove all over the city. These streets were just as overcrowded with people, dogs, monkeys, and cows, all traversing past the brightly lit shops and eateries. During this time, I did some basic HTA touch energy therapy with both Jai and myself. Compared to me, she was calm and very easy going, willingly nestled between my legs and sitting there with ease as we nimbly zipped through the streets. My intention with the energy work was to prepare us both for this new unknown journey together. With one hand placed on her heart Chakra and the other stroking her head and upper back, I felt my tension subsiding and I smiled, knowing we were meant to be together and would share beautiful moments in the future.

It took us several tries, turns, and stops to ask people where to go, but we found the vet clinic. When we finally got there, it was somewhere in a rundown, brightly lit and tiny room behind a rather small fast food joint that was surrounded by tall apartment homes. I was greeted by the vet tech, and about a half an hour later, the vet, who was on her way out of town, stopped by for only a few minutes. She just looked at Jai briefly, saying that she looked good and seemed healthy. She gave her a rabies shot and some pills for deworming and left. It didn't seem like a complete health exam to me, nor did it seem like I got anywhere even close to understanding what the next steps would be to prepare for Jai's adoption. I had expected to discuss the international transportation plan, learn about the protocol, and get the right paperwork together by the beginning of the following week so I could visit a government office and get official transportation approval. I would also need to get all the travel supplies in place and would have to make sure the airlines allowed her to be transported with me on the same flight itinerary. I was disappointed when this was not the case and I wasn't given an opportunity to get the information I needed. My heightened state of adrenaline had dropped somewhat, and I realized I was going to need to do massive research and contact a lot of people in order to safely transport Jai out of India.

Since I was not able to keep Jai with me in the hotel room and it was getting late, we needed to get Jai to the boarding facility for the week. Jai was going to stay with a dog trainer at his family home, which was located in the southern part of town. It would take another 30 minutes to get there, and it was nearing 11 p.m. His home was not easy to find, and Roop needed to call him several times in order to have him lead us across a busy highway to a little side street where he lived and boarded dogs. Thankfully, I was able to board Jai there for the week until I could figure out how to transport her to the USA. His family was kind and supportive, which made me feel comfortable leaving Jai behind.

Jai

## *Travel Arrangements for Jai*

Throughout the week, there was a lot of planning and preparing for Jai's trip to the USA. I chose a company named Petfly that was referred to me by United Airlines to transport Jai from Delhi. I would need to rent a driver to transport Jai up to Delhi, a four-hour, one-way car ride on the following Saturday, the day prior to my departure. Petfly is a well-organized and highly-recognized professional organization run by a team of veterinarians located in Delhi in Northern India. They have been instrumental in transforming and evolving the global concept of international animal rescue and safe international transportation for pets. They have created a solid network across the globe, and they arrange for everything needed, including vaccinations, health certificates, government-approved documentation, and all the travel necessities in order to transport pets safely to their new destinations. Their network of volunteers is vast; some will travel with animals directly in the cabin, or they make the arrangements and place the animal directly into a pressure-controlled space in the cargo area of a plane for safe transport.

Bee, a cousin of Sunil, who is a good friend of mine, loved to drive and would be happy to drive us to Delhi. Since I only had one week to get everything arranged for Jai, I was happy when I received word on Friday night that Jai was cleared to travel to the USA the following evening on a redeye. The following morning, Bee met me at the hotel at 9 a.m. for my next adventure with Jai. He patiently helped me make several stops to retrieve Jai before embarking on the long trip to Delhi. The stops went relatively smoothly, but they took more time than I had anticipated, as they tend to do in India. When we finally had Jai in our possession, I was so relieved and excited to spend time with her in the car. This was probably her first car ride, and I would be able to do more HTA energy therapy with her to customize her to the car ride and get her ready for her vet visit, bath, and long plane journey that evening.

Sitting in the back of the car with Jai curled up next to me, I looked at her with disbelief. How had I been this lucky to have found such an incredible sweetheart? I felt so relieved and at peace to have her by my side. No doubt, she would teach me a lot over the years to come as she became part of my family. She had an independent, but ever so sweet, personality, and I was still unable to completely grasp that she was really coming home with me.

Smiling inside and out with pure happiness, I sat back for the next several hours, letting the warm wind that blew through the open windows wash over my soul, as I watched the countryside pass and reminisced about the unbelievable month I had spent in beautiful India. With all the hard work to get to this point now behind me, I had to let go and trust that this next portion of the adventure would run smoothly and without glitches. We were almost at our final destination.

## HTA Energy Therapy with Jai

During the car ride, I used two steadfast techniques with Jai, Vibrational Grooming and Bridging. Through my loving intentions and the love shared between us, I opened up the energy channels to the Divine and Mother Earth, our ultimate energy source. I visualized and channeled directly that Jai would travel without stress or anxiety and would settle into her long journey to the USA. In the beginning of the car ride, she was drooling and panting profusely, signs that she was obviously nervous. I held her gently and stroked her throughout the whole car ride. The shift in her was easy to notice as she started to settle in and stretch out, while taking some deep breaths. She would snuggle into me and close her eyes for a nap. Her instinct correctly sent the signals that she was safe with me and could release and relax fully. We stopped in the middle of our journey for a little chai and took Jai for a walk in the countryside. By this time, I had a proper leash and collar for her. Jai was so easygoing, walking nicely on the leash as she sniffed, smelled, and did her business. Being out of the city, there was less chaos, and the quiet peace was a relief to the both of us.

Back in the car, we enjoyed more snuggling, touch therapy, and relaxation together until we arrived in Delhi around 3 p.m.

## Delhi

We made it to Delhi in a little over three hours, where Dr. Premlata met us. Tears started to roll down my face as we hugged with relief and trust. Beyond a doubt, I had made the right decision to drive Jai up to meet her. We had many phone conversations during the past week, and I wanted to see her operations and connect with her in person. Her clinic is clean, and her staff is very professional.

Through my experience in India, I know that promises fall short, and people say they will help you but ultimately do not. This is frustrating, and I believe it is due to massive cultural differences, which I now am more accustomed to. Meeting Dr. Premlata, however, was a relief, and I felt blessed and happy that she would take control until Jai's arrival in the USA.

I finally felt that I was able to breathe again and that all of our efforts would be rewarded.

Thinking back to a conversation I had with my husband earlier that week, I smiled. I had been in despair, talking about the huge undertaking of getting Jai out of the country and through customs in the USA. He has always been such a positive and rational soundboard, and he simply said, "You, of all the people I know, will succeed in bringing her back here." He obviously knows me rather well; at this point, it sure did look like we were going to succeed in this mission.

Jai was examined, given all the necessary vaccines, and taken for a bath to prepare for transport that evening. Dr. Premlata's impeccable reputation with the government allowed their approval paperwork to be expedited. Jai would arrive a few hours after my arrival in San Francisco on Monday morning. I was bringing my Jai, another soul mate, home with me to integrate into my family and her soon-to-be family: her human father, my husband, Tom, and our three other rescue dogs, Kaya, Kole, and Copie.

## Leaving Jaipur and Reflections

After a long and grateful goodbye with Dr. Premlata, I climbed into the passenger seat next to Bee. I was exhilarated and exhausted at the same time. Finally, during the three-plus hour trip back to Jaipur, I was able to fully release, and my tears would not stop. Again with the windows down, I simply stared out at the passing countryside. I then closed my eyes, letting the warm evening air travel over my face and blow through my hair. It was almost 6 p.m., and our mission was accomplished. Now all we needed to do was drive back to Jaipur. But there was more for me to do. I was due to fly out the following day at noon and still needed to pack up and say goodbye to my dear friends.

With my eyes closed and tears streaming down my face, I could feel that it would take a while for my energetic body to recalibrate back into a steady rhythm. My soul and inner being had not caught up to the tremendous actions accomplished that day and during the past month. This whole journey had been a success in so many ways and a turning point in my life. This final day and car ride with Jai, with our constant touch and bonding, was just the icing on the cake.

When I opened my eyes, I was mesmerized as I looked right at the most intense, gigantic, golden and orange evening Indian sun. It was the biggest I had ever seen. The incredible sunset brought my chapter there to a close. With this representation of nature, it was quite obvious that the Divine was smiling down to me. What I had come to do had been accomplished, and it was time for me to go home. A lot of new doors were about to open up for me, and I sensed this message loud and clear. What an incredible day it had been and what a perfect end to my adventurous bold and solo journey to India.

That evening, in the quiet of my hotel room, I continued to recall all the accomplishments during this past month. I had pushed myself to limits I did not know I possessed, and I had grown and changed exponentially because of that. I would carry forth, traveling through life with higher vibrations, being more vulnerable and open with each step, and being brave to live big. I was returning to the USA with an intensified purpose to my life. I could tell there would be many changes in the coming year, and a whole new chapter was about to start. I was ready and at peace with all of the unknowns, blessings, and magnificence of life. My first research study with elephants was complete, my impromptu case studies with Bhitily and goats were complete, and bringing Jai home with me was also well underway. There was nothing else I wanted or needed at this juncture; I felt complete.

The next day, I left for the USA, and only a few hours after my arrival in San Francisco, I was able to bring Jai through customs as a new citizen of the USA. My new little girl had endured the long trip, and I was excited to introduce her to the rest of the household.

## Integrating Jai in California

I thought the integration process was going to be easy with Jai and the other dogs, but surprisingly it was a lot tougher than I had expected. Jai's sweet temperament and the fact that she had been surrounded by a hundred street dogs and so many people in India made me assume that she would get along well with anyone, including my family of dogs. I had no premonition that integrating her into my dog pack of three other rescue dogs would actually be quite difficult and a project in itself. As it turned out, she was immediately extremely social and gentle with humans and even children, but not with the other dogs. She actually was extremely territorial and vicious in certain instances. In the beginning, I had to separate the dogs and slowly start integrating them, keeping them at a safe distance from one another. During the first few weeks, Jai spend a lot of time with me in the car, which was a safe place for her and where we were able to build a stronger understanding of one another.

As the weeks progressed and with my consistent training efforts, she started to settle into her new routine and was becoming more comfortable with the other dogs in the house. I knew there were shifts taking place since she did not exhibit signs of aggression when we were all on hikes together. I knew that, in time and with the proper positive guidance, she would come to accept the other dogs and wouldn't try to attack them. I saw that shaping her environment with care, providing solid boundaries, and being a role model would allow her to become more confident and secure and she'd learn to deeply trust. From experience, anyone's behaviors are adjustable when given a positive environment. The key was to train by using positive reinforcement, without forcing anything, and to listen and observe in order to preserve the trust we were building. I believe in relationship building, and trust is a key component of that.

I have been training many dogs over the years and knew it was my role to assist Jai through this major adjustment and keep everyone physically and emotionally safe. I believed that the four dogs would get along and become friends in due time.

Even though there were lots of little positive movements of change taking place, it took about three months to finally see things align on a bigger scale. Almost a year after her adoption, we are about 90 percent there. As a family, we have gone through a lot of adjustments, and we have done a lot of teamwork to integrate Jai and structure the household in such a way to create a successful environment where everyone will thrive. It takes a communal effort.

I continue to integrate energy therapy practices, such as touch, sound therapy, and the use of Rescue remedy and other Bach Flower essences, EO's, and CBD oil with the dogs to facilitate balance and a higher energetic frequency inside their bodies. I use behavior and training techniques and add in exercise and outdoor adventures to stimulate them physically. I integrate all of these components on a daily basis to promote consistent and positive change for our family, as well as in the world with other dogs outside the home, enhancing the positive relationships between all of us together. I allow for this connection to run its natural course.

Jai needs time to learn that she is surrounded with abundance in her new environment and does not need to react based on her prior survival street fears. There is no lack of anything in her new world. She is safe and has a loving and caring family. Through combining and applying these positive concepts and therapies together in a holistic fashion, I have seen such tremendous progress.

Every day, when I observe Jai as she sits on the deck overlooking the marina, the vast San Francisco Bay, and Mount Tam, I know she is taking in this different kind of beauty and is at peace in her new home. I feel more connected with her daily. She finally knows that her name is Jai and chooses more often than not to listen to my call. Even though we do not speak the same language, I only need to look at her to know that we are one and the same. She will sit for hours watching the sunrise and sunset, and she seems lost in thought. Perhaps she is reflecting on her journey or is merely taking in the beauty that surrounds her in her new home. I also hope that she is able to reflect back to where she came from and the different chaotic beauty that surrounded her in India in her prior days. Many say that animals can, indeed, perceive different forms of beauty, and I believe that all of us are affected by the environment we live in. If it is a peaceful and pleasant environment, we have a chance to heal from our past traumas or mishaps that may have shaped our earlier years. Living in a wholesome environment allows the ability for all species to live more balanced, longer, happier, and healthier lives.

## *My Observations*

I am always reminded and ever so grateful that I am actually the one that learns so much from all the animals that surround me daily. They help me to be more patient, listen more deeply, and be more understanding as I become a better animal guardian. My own dogs encourage me to continuously seek out other animals in need of assistance, and they bless my path. They show me to be bold and brave and to listen to and trust my instinct through Divine guidance, intuition, and the powers of the universe. There is no doubt as the days pass that it was my purpose to adopt Jai, just like all the other doggies that have crossed my doorstep. In doing so, my knowledge, understanding, and compassion for them and other animal species has elevated exponentially. My work with animals has become more profound every day.

# Chapter 10

# INSIGHTS AND STEWARDSHIP

*Humanity must become aware that if we take away the Beautiful beings of nature, you will take the Beauty out of the very things that make the world and our lives Beautiful. Our soul and spirit will become incomplete.*

~ Anonymous

Our modern world has validated that animals have a conscious mind and, therefore, have emotions and feelings. They create memories, have high levels of intellect, are social beings able to interact with other species, and have incredible communication skills.

It is becoming quite clear that our human role as being "superior" needs a recalibration, and we need to learn how to listen to animals and adjust our moral compass to be leaders and stewards of their well-being, happiness, and survival.

Scientists need to share their findings with nonscientists, and we all need to listen and validate the knowledge of people who have nontraditional views, such as the Mahouts and Indigenous people who have lived in close connection with wild animals and nature.

"While we live in a scientific world, we cannot do it alone. We need help from other disciplines, including those whose consensus are with theology and spirituality. The 'hard' and 'soft' sciences can be unified to produce 'deep science'... Hard science, socially responsible science, compassion, heart and love can be blended into a productive recipe for learning more about the fascinating lives of other animals and the world within which each of us lives... Animals depend on humans to have their best interest in mind. We can choose to be intrusive, abusive, or compassionate (Bekoff 2002, 30)."

Let's work together with brilliant pioneers in this vast animal field, such as Jane Goodall, Cynthia Moss, Carl Safina, Jeffery Moussieff Masson, Kathy Payne, Lawrence Anthony, Marc Bekoff, Joyce Poole, Joe Dispenza, Frans de Waal, Charles Darwin, Larry Brilliant, Robert Greene, Carol Komitor, Amy Snow, Nancy Zidonis, Carol Gurney, Martha Williams, Temple Grandin, Vicki Croke, Paul Barton, national and international animal rights groups, NGO's, and many more who work diligently to better the lives of animals every day and deserve credit.

I see the benefits and effects of this work from my time spent being an HTA certified practitioner and conducting this work in California and in other parts of the world. Traveling to India, filming HTA animal energy sessions, and gaining solid findings through my research, I am driven to continuing my learning, conducting more research in this expansive field, and publishing my findings.

We all need to be stewards and do our part, taking a higher interest in preserving the planet and all the animals that reside here. By learning from others and educating ourselves, we are able to understand on deeper levels the true nature and incredible abilities that all of Earth's species possess. We share this incredible planet with so much plant life and living species, and we should preserve our intertwined ecosystem. As we evolve, our kinship needs to connect with more depth. We are smart enough to behave with high moral standards, respecting and treating all forms of life with appreciation and care. We should be able to preserve and save species from extinction and step away from cruelty and destruction, as we are all citizens of this earth.

Animal energy therapy modalities are easily accessible around the globe, and their integration has such tremendous benefits. Through these modalities, not only will we attain a greater under-standing with other species, but I believe we will be able to heal the emotional wounds that many have received through prior human domination and violence. These animals will be able to move forward, release past traumas, and create new healthy relationships with humans. Through the integration of energy therapy practices, we are able to use touch, create beautiful sounds, and add natural scents to enhance the lives of our animals and ourselves. We have the ability to learn any style of energy therapy practices ourselves, or we can hire certified professionals to help take care of our animals. Animal energy therapy is a valuable modality of healing and recalibrating the body, one that we can all integrate easily, knowing its tremendous benefits.

## Insights from Others

Throughout history, wildlife, including elephants, have always been of high interest, and humans have developed strong bonds with elephants. As Marc Bekoff states, "It is because animals have emotions that we are drawn to them. We need animals in our lives, animals are sentient beings who experience, the ups and downs of daily life, and we must respect this when we interact with them (2002, xxi)." Love is often thought of and spoken about by many as being the solution for almost everything. "We must intimately connect with and love other animals, other humans, and all environments if we are to continue to live with grace and in harmony on this wondrous and interconnected planet. We need animals, and we need wildness and wilderness, to be healthy human beings (2002, xxi)."

## Stewardship

I share Carol Komitor's belief that heightened awareness of animal energy therapy will lead to its availability in every household. Together, we should encourage others to increase their hunger for education about animals and advocate for animal rights. Each of us has the opportunity to push for policy and govern locally or globally.

Being human calls us to inspire each and every one of us to share, give back, donate our time, and do whatever it takes to protect and care for the planet. We must lead by example and operate with integrity and authenticity, and through higher vibrations, moral standards, and ethical practices that will benefit and not harm living beings and organisms. We need our full-range ecosystem to flourish if we want to continue to survive as a human race. If we take care of all species on our planet, we will help ourselves and the children who will inherit this world. It is my hope that we are all inspired and will want to increase our own efforts. If we listen to the universal call to create stronger communities, to share and give to one another and all species, and treat others like we want to be treated, the universe will provide us all with immeasurable benefits in return.

# WORK WITH ANNE-FRANS

Anne-Frans continues to promote her work through speaking engagements and learning opportunities. She works with others to facilitate and integrate animal energy therapy practices and conducts HTA energy therapy sessions and behavior and training modification practices across the globe. She continues to conduct her research studies with different species to promote alternative healing and energy therapy practices. To hire Anne-Frans as a researcher, animal energy therapist/facilitator, or keynote speaker at your next event email annefrans@annefrans.com.